T0270925

ASTROPHYSICAL JETS AND BEAMS

Astrophysical jets are spectacular displays of gas or dust ejected from a range of cosmic bodies; they are seemingly ubiquitous on scales from comets to black holes. This volume reviews our understanding of jet processes and provides a modern guide to their observations and the role they play in many long-standing problems in astrophysics. It covers the major discoveries in gamma-ray bursts, solar and stellar jets and cometary jets. Specific physical processes for all classes of jet are illustrated and discussed in depth, as a backdrop to explaining spectacular jet images. Current jet models raise as many issues as they solve, so the final chapter looks at the new questions to be answered.

Written at an entry level for postgraduate students, this volume incorporates introductions to all the governing physics, providing a comprehensive and insightful guide to the study of jets for researchers across all branches of astrophysics.

MICHAEL D. SMITH was awarded his Ph.D. in astrophysics by the University of Oxford in 1979. He is now the Director of the Centre for Astrophysics and Planetary Science, Director of the Kent Space School, Director of Research for SEPnet (South-East Physics Network), and holds the posts of Director of Graduate Studies and Sub-Dean in the Faculty of Sciences at the University of Kent. He is a member of the International Astronomical Union and a Fellow of the Royal Astronomical Society.

Cambridge Astrophysics Series

Series editors:

Andrew King, Douglas Lin, Stephen Maran, Jim Pringle and Martin Ward

Title available in the series

ASTROPHYSICAL JETS AND BEAMS

MICHAEL D. SMITH
University of Kent, Canterbury

CAMBRIDGE
UNIVERSITY PRESS

CAMBRIDGE
UNIVERSITY PRESS

Shaftesbury Road, Cambridge CB2 8EA, United Kingdom

One Liberty Plaza, 20th Floor, New York, NY 10006, USA

477 Williamstown Road, Port Melbourne, VIC 3207, Australia

314–321, 3rd Floor, Plot 3, Splendor Forum, Jasola District Centre, New Delhi – 110025, India

103 Penang Road, #05–06/07, Visioncrest Commercial, Singapore 238467

Cambridge University Press is part of Cambridge University Press & Assessment, a department of the University of Cambridge.

We share the University's mission to contribute to society through the pursuit of education, learning and research at the highest international levels of excellence.

www.cambridge.org
Information on this title: www.cambridge.org/9780521834766

© M. D. Smith 2012

First published 2012

A catalogue record for this publication is available from the British Library

Library of Congress Cataloging-in-Publication data
Smith, Michael D. (Michael David), 1955–
    Astrophysical jets and beams / Michael D. Smith, University of Kent, Canterbury.
      p. cm. – (Cambridge astrophysics ; 49)
    Includes bibliographical references and index.
    ISBN 978-0-521-83476-6 (hardback)
    1. Astrophysical jets. I. Title.
    QB466.J46S65 2012
    523–dc23   2011047479

ISBN   978-0-521-83476-6   Hardback

A dedication to my wife, Daniela,
to whom I owe the leaping delight

Contents

Preface

Jets are amongst the most mysterious phenomena to be discovered in modern astronomy. They are able to form and propagate under almost all conditions associated with a vast range of astrophysical objects. This book is concerned with all the diverse jets which have so far been found beyond our own planet. It will be seen that our universe is replete with jets because they act as essential outlets or valves for regulating the birth and early development of discrete objects and their extended environments.

The purpose here is to assimilate all we know from the different disciplines in which they are encountered. I cannot try to review radio galaxies, star formation, comets or planetary nebula, but only the parts in which jets are essential to their understanding. We thus learn about the driving mechanisms involved and their consequent impact, and so learn to appreciate the diversity. The idea is to accumulate, perhaps possible for the last time, all the material which relates to *the* phenomenom referred to as jets. Hence this is not a series of reviews but a gathering of essential knowledge. And, consequently, by establishing their common properties, this book hopes to represent a turning point in what we have come to understood as jets and what we will go on to discover.

It will be attempted to make this book self-contained with a modicum of required knowledge. It should serve as a timely introduction for astronomy students who seek to develop a broad approach to understand the 'bigger picture'. On the other hand, the theoretician may relish the range of phenomena which depend upon supersonic flow and shock waves for their explanation. In my early career, my research focused on the extragalactic variety before veering towards the emanations from young stars and protostars. In this time, the astrophysical jet has seen only rare complete reviews. It has been twenty years since the last comprehensive jet book and the subject has moved on. Even in the last five years, since this book was started, tremendous progress has been witnessed with wide-field astronomy, serendipitous discovery and space rendevous providing new types of jet with data that often contradict our preconceptions and challenge our conceptual skills.

What knowledge is essential to understand jets? It would be inexcusably naive to suggest that just a few physical processes lie at the heart of the matter. We will have to become familiar with a number of launching configurations, radiation mechanisms and observing techniques. Yet, there are common threads to bind the book's material: above all, gas dynamics – especially the behaviour of gas accelerated from low speed to high speed. A second strand is the cause of the containment or collimation of this flow. Thirdly, the impact as the high-speed jet is disrupted or abruptly terminated. These flows invariably involve shock waves, the sudden transitions within a jet driven up to supersonic speeds. Hence, some general insight into fluid flow or magnetohydrodynamics is an advantage, while some understanding of the underlying equations must be developed as our intuitive understanding of flow patterns becomes insufficient.

I have taken pains to ensure that this book contains a useful guide to the literature by including a sensible proportion of citations to research papers. However, it is not a reference book and many authors in the field will be disappointed to find that their names and their pet jets are not directly linked. My apologies. My hope is that the papers which are cited can be consulted not as the authority, but as apt starting points for forward and backward panning through the archives. Without this strategy, the task of presenting a concept such as the astrophysical jet within one medium-sized book would be unwieldy.

The *Astronomical Journal*, *Astronomy & Astrophysics*, the *Astrophysical Journal* and *Monthly Notices* are the four time capsules supporting the cornerstones of this work. It would, in addition, take a separate volume to thank the individuals who have contributed to the knowledge within. I will therefore limit myself to thanking all my colleagues in the School of Physical Sciences and all my collaborators in the jet set. The work of several communities of jet setters is embodied here; all I have done is to take their treasures and, hopefully, reveal them. However, there remains one treasure more mysterious and alluring – still as true as in 1979, for inspiration I thank Daniela Rohr.

I

Introduction

1.1 Rudimentary definitions and concepts

A man-made jet is a narrow stream forced out of a designed aperture or nozzle. Water fountains and jet engines provide everyday examples of liquid and gas jets. Skin penetration and rock drilling are high-technology applications. Jets also occur naturally on the Earth associated with geysers and some types of volcanic eruption. These terrestrial jets arise when material is raised to a high pressure below the surface and is forced to ascend through channels with rigid walls. In contrast, the astrophysical jet involves relatively unfamiliar physics, usually under extreme, but occasionally in exotic, conditions.

In astronomy, there is rarely a solid nozzle or tube to align the jet flow. The material is driven, with a few exceptions, through an interacting gas. In other words, an astrophysical jet is a slender channel of high-speed gas propagating through a gaseous environment. The exceptional jets are the nearest extraterrestrial jets associated with comets such as Hale-Bopp. In the latter case, as for those shown in Fig. 1.1, they are believed to form when a high-pressure mixture of gas and dust breaks through vents in a solid crust.

Astrophysical jets are driven from diverse objects on very different size and mass scales. They can be produced from the vicinity of supermassive black holes in the case of active galactic nuclei (AGN), by star-sized black holes in microquasars, by neutron stars in some X-ray binaries, by protostellar cores in young stellar objects, and by white dwarfs in symbiotic binaries and supersoft X-ray sources. The 'material' of these astrophysical jets is much more than a simple compressible fluid or gas. The gas may consist of a mixture of ions, electrons, molecules and dust particles, or can be dominated by a magnetic field and relativistic particles. The complete quantitative inventory has proved remarkably difficult to establish in all cases.

The truly remarkable fact is that, despite a lack of rigidity, the materials and forces conspire to generate jets with high thrust and power from all these astrophysical objects. The thrust is often sufficient for the jet to excavate a tunnel which transports the gas tremendous distances. For example, jets from the vicinity of supermassive black holes, located deep in the nuclei of galaxies, pierce through the interstellar medium and exit into the extragalactic medium (Fig. 1.2). The same jet is still operating, at least in the same direction, despite ten orders of magnitude (ten powers of ten) expansion. Even so, the jet phenomenon was not deemed to be reason enough by some to deserve much attention. After all, aren't these merely waste channels – the rising smoke from the chimney pots? By studying the smoke, we could never hope to discover the cause of a fire.

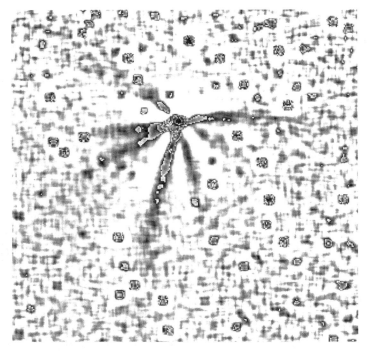

Fig. 1.1 Seven jets emanating from comet Hale-Bopp observed on 18 August 1996. The jets are detected through the reflection of sunlight by dust grains. The dust is ejected from vents on the surface due to the pressure of out-streaming gas. The 20-second CCD image with the R-filter to measure the dust has been heavily processed to suppress the smooth structure of the coma. At the time of these observations, comet Hale-Bopp was 2.761 AU from the Earth and 3.392 AU from the Sun. The image size is $320'' \times 320''$. Credit: N. Thomas, H. Rauer, Danish 1.54-m telescope at La Silla, ESO.

That opinion has turned out to be a misconception. Instead, the jets must be a crucial component of our observational and theoretical interpretation. Firstly, the mass and power ejected are often significant fractions of that available. Therefore, how we trace the evolution of the central engine must be considerably modified. Secondly, the jets are now believed to extract the angular momentum from the inner contracting or collapsing zone – without the jets, centrifugal forces might suffice to support material within a spinning disc. As a consequence, a collapsed object would not so readily form.

The jets are also our beacons. Entire classes of distant objects in the universe have been made detectable through their jet activity: their luminosity is boosted by relativistic beaming for those members with jets pointing by chance in our direction. Without the jets, few or none of the objects would be powerful enough to be detected.

Even within our own galaxy, we discover and precisely locate stellar embryos through their jets. Jets from protostars, the contracting cores of gas which will eventually form the stars, cut through the thick obscuring cores and plough on through embedding giant molecular clouds. The protostar is visually obscured and their jets would also be hidden, since stars are born deep inside dense cores. Through impact with a less obscuring molecular

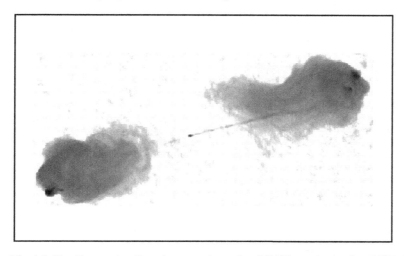

Fig. 1.2 The Cygnus A radio galaxy, catalogued as 3C 405, as observed at 5 GHz in the radio by Perley *et al.* (1984) with the Very Large Array (VLA). The resolution is 0.4". It is the closest example of a powerful double radio galaxy, identified with a cD (giant elliptical) galaxy at a redshift of $z = 0.0561$ corresponding to an angular size distance of 220 Mpc. Two kiloparsec-sized hot spots are found at the leading edge of each 60-kiloparsec lobe. Reproduced by permission of the AAS.

cloud after penetrating out of the core, molecules are heated and subsequently emit radiation at wavelengths longer than in the visible. This infrared, millimetre and radio emission is less attenuated by intervening dust grains and so a jet makes its presence known (Fig. 1.3).

Jets also provide a channel for feedback. From the cores of galaxies, they may support and sustain gas in a halo and so slow down the rate at which stars form. They may also trigger bursts of star formation in cloud complexes close to their paths. From the cores harbouring protostars, jets raise the pressure upon impact. The pressure lowers the critical mass required for surrounding clumps to gravitationally collapse. Moreover, the jet-incurred turbulence speeds up the star formation 'metabolism', enhancing both dispersal and collapse.

Besides these motivations, jets are widely exploited as laboratories within which dynamical, physical and chemical processes are examined in concentrated form. The focusing and compression lead to rapid reactions: strong heating in shock waves and swift cooling in their wakes. The product is bright compact knots of emission. Consequently, they are testing grounds in which many predictions have been verified and others forgotten.

In this introduction, we piece together the story of the discovery of jets in what has assembled into an increasingly divergent range of phenomena. This motivates the major aim of this text: to raise and study the concept of *the astrophysical jet*. Independently of specific causes and effects, jets are spectacular in display and fascinating in properties. We study them here as flow systems, searching for their common attributes and unifying phenomena.

Our definition of a jet is quite narrow and exclusive. Beams of radiation (light rays) are obviously excluded. Plumes of gas are also excluded. In fluid dynamics, plumes are features maintained by thermal buoyancy. In astrophysics, plumes refer to wide spouts which open

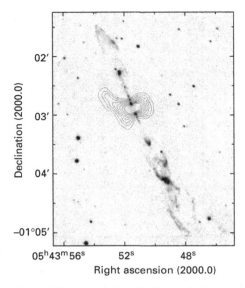

Fig. 1.3 The symmetric HH 212 outflow driven by twin jets from a concealed, centrally located protostar. It is located in the L 1630 molecular cloud about 90″ north-east of the Horsehead nebula at a distance of ∼460 pc. The outflow is shown in emission from vibrating molecules of hydrogen (H_2 1–0 S(1) at 2.12 μm). The contours correspond to emission levels from ammonia molecules (integrated NH_3 (1,1) emission at 23.7 GHz), produced from cold, dense gas. This molecular core harbours a protostar detected in the far-infrared. Credit: Wiseman *et al.* (2001), Calar Alto & Very Large Array (VLA). Reproduced by permission of the AAS.

up, i.e. the fluid follows divergent streamlines. Hence, they may well be associated with deceleration or instability. It still remains to draw a line between a collimated wind and a jet. For example, a length twice as long as the width sufficed in the past for a milli-arcsecond quasar feature to be described as a jet. These jets were observed with the continent-wide VLBI – the Very Large Baseline Interferometer. A minimum ratio of 4 has also been adopted by some astronomers.

The semantic distinction between jets and beams of plasma has changed through the years. Originally, jets were observed and beams were the hypothesised underlying flows. Common usage, however, has now evolved so that jets are both observed and theoretical. We will save the terms beam and beaming to refer to the relativistic effects caused by jet flows at speeds close to the speed of light. In this case, beaming refers to the anisotropic distribution of radiation that is produced in the reference frame of the central ejecting object. While necessary, these definitions may make better sense after we consider how the jet concept entered the daily working lives of many astrophysicists.

1.2   Jet presence and function

Our definition of a jet is not limited to those we can see: the energy-dissipating visible variety. In the past, even though the evidence for underlying jets has not always been compelling, the assertions have subsequently proved remarkably accurate.

There are strong reasons to believe that the production of jets is related to the efficient removal of angular momentum from accreting or collapsing systems. If they do their job efficiently, they allow almost all the spin to be extracted by ejecting a small fraction of the available mass. A high-performance jet would also be made to transport the energy a long distance with little loss on the way. Ironically, the best jets are then invisible. With no dissipation there is no radiation, and detection may then rely on reflection or absorption of external sources of light. In such cases, their presence can only be inferred.

On the other hand, there are reasons why jets could reveal their presence in all three fundamental flow phases. Firstly, at the point of launch, the driving mechanism is unlikely to be 100% efficient at converting energy into bulk flow. In the conversion process, some energy is dissipated into random particle motions which may lead to considerable radiative losses. Secondly, while propagating, velocity structure may grow within the jet flow. Velocity gradients tend to steepen into internal shock waves, and turbulence is generated by velocity shear or magnetic reconnection. Both shock waves and turbulence convert kinetic energy associated with coherent large-scale motions into thermal energy.

Finally, the interface with the ambient medium often provides indirect evidence and can be mistaken for the jet itself. For example, jets may excite a surrounding sheath of stationary gas or they may continuously supply an active region at the point of termination. This impact region is called a 'hot spot' in extragalactic terminology and a 'bow shock' in the star formation context (with those detected in optical light termed Herbig–Haro objects). Both are produced by shock waves which are by nature both dissipative and compressive; since radiation processes are sensitive to wave excitation and collision rates, these are ideal regions for strong emission.

An underlying jet may also be inferred from the detection of a few separated knots of emission which are well aligned. However, some evidence which suggests that they are being more or less continuously energised is required. This property distinguishes a jet flow from a ballistic flow of independent clumps of gas – the proposed 'plasmons' of radio galaxies and the 'bullets' of stellar jets.

The conclusion is that there are various conceivable efficiency coefficients for jets. We can define a transport efficiency in terms of the ability to transfer momentum or energy. We can define a propagation efficiency in terms of the ability to tunnel through the ambient medium. Finally, we can define a spin efficiency in terms of the fraction of the inflowing angular momentum which is extracted.

1.3   Early history

For centuries, jets of a kind have been discussed in astronomy. These had been observed from comets. Brilliant streams had been recorded, emanating from the nuclei of comets and undergoing sudden extensive changes. As far back as 1682, a distinctive curved jet appears in a drawing of Comet Halley by Hevelius (Hevelius, 1682). In 1836, Friedrich Bessel recorded 'out-streaming' from Comet Halley over arcs of 30–90°, rotating over hours and days on the sun-facing side (Bessel, 1836). He developed a theory to account for the brightness of the comet based on the evaporation from a solid nucleus. He realised that jets had an important consequence: a rocket effect. Material evaporated from the surface of the nucleus exposed to the Sun could provide the non-gravitational forces which alter the orbit of Comet Encke. A jet propulsion effect was thus thought to influence comet trajectories.

In 1881, Captain Noble remarked upon apparent spouting from the nucleus of Comet b, 1881 (Noble, 1881). In 1882, the rocket effect of a jet on this comet was discussed by C.E. Burton (Burton, 1882) as leading to non-gravitational disturbances. Specifically, he expected motion transverse to the orbit, rather than retardation or acceleration, to be detectable. In 1950, Whipple confirmed this with detailed calculations (Whipple, 1950) but the jet model was not developed. These days, jets are considered responsible for supplying the entire coma. They can be highly collimated, as we will discuss in Chapter 8. Only very recently, however, have we been able to approach comets and locate the source of the jets.

Meanwhile, in 1918, an extragalactic jet had been discovered by the astronomer Heber Doust Curtis at Lick Observatory (Curtis, 1918). It was reported as a 'curious straight ray'. As shown here in the later image of Fig. 1.4, it is 'apparently connected to the nucleus' of the nebula Messier 87 (M 87), also catalogued as NGC 4486. At that time, the nature of the ray as well as that of the nebula were unknown. With no basis, the phenomenon remained undiscussed for over forty years. Incidentally, Curtis had previously been aiding Wilson on recording the Halley jets (Wilson, 1910) but refrained from employing that description in his own work.

In 1954, Baade and Minkowski investigated the curious M 87 feature and renamed it as a 'jet' simply on a hunch that the feature could be part of a rapid outflow (Baade & Minkowski, 1954). They actually had some circumstantial evidence since, although the spectrum of the jet revealed a featureless blue continuum, an outflow had been associated with the innermost regions of the galaxy. The outflow was identified by a strong oxygen

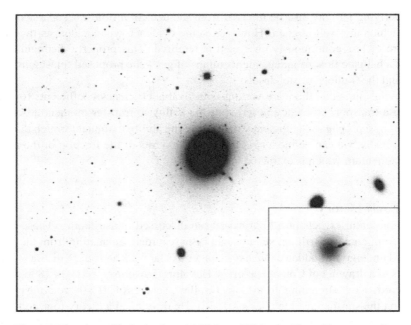

Fig. 1.4 The giant elliptical galaxy M 87 (type E0) in the Virgo Cluster at a distance of 16 Mpc, as imaged by the Kitt Peak 4-metre Mayall telescope in 1975. The image displays a 25" sub-cluster. The peculiar jet emanating from the core of M 87 is only visible in the processed image shown in the inset 1.4" panel. Credit: NOAO/AURA/NSF.

emission line, blue-shifted by 300 km s^{-1} relative to the galaxy's nucleus. As it turns out, this is a thousand times slower than the jet as presently constrained.

Baade and Minkowski recognised that the jet could be just the visible apparition of a 'peculiar condition' extending over an enormous volume. This volume is occupied by material that generates radio emission. In fact, M 87 was associated with Virgo A, the third-strongest radio source in the northern sky. It was, hence, baptised as the Crab Nebula of Active Galaxies and it was reasonable to suppose that, like the amorphous Crab, the optical jet as well as the newly discovered radio galaxies were being observed through radiation produced by the synchrotron process (see § 1.6). Synchrotron radiation is emitted when a relativistic electron spirals around magnetic field lines. In 1955, Shklovskii made the critical prediction that the radiation should then be strongly polarised, which was soon verified by Baade (Baade, 1956).

As other jets were discovered in radio galaxies and quasars, such as 3C273, 3C48, 3C 196 and 3C 279, the M 87 jet became the classical example of the phenomenon, being so nearby and bright. Quite peculiar was that these first jets were all one-sided, whereas almost all radio sources were double. A second early problem was that the relativistic electrons should lose their energy well before propagating down the jet from the nucleus. Hence, the jets should not remain visible for such lengths (Shklovskii, 1963).

The discovery of the powerful extragalactic radio sources led to the need for an energy supply, which had to lie in the quasar or galactic nucleus. The typical radio source consists of two lobes symmetrically located about a central galaxy (seen at optical wavelengths). After the identification of many radio sources with galaxies in 1949, the first radio galaxy found to be bisected was Cygnus A in 1954. After this, the third Cambridge (3C) catalogue, released in 1959, contained images of many of the brightest radio sources and established prototypes for many of the categories now used as household names by jet enthusiasts. Subsequently, various catalogues and classes of radio sources, associated with jets launched from diverse types of galactic nuclei, have been discovered. These will be explored in Chapters 4 and 5.

The idea that the supply might be continuous from within a parsec out to hundreds of kiloparsecs was not quick to emerge. Jets were rarely detected and twin jets not at all envisaged. Therefore, interpretations in terms of non-continuous ejections were favoured. Scenarios in which clouds of plasma, later packaged as 'plasmons', were somehow breaking out along an axis of least resistance (Shklovskii, 1963), were developed. A gravitational slingshot mechanism was also proposed (Saslaw *et al.*, 1974). The advantages of expelling relativistic gas and magnetic field in clouds later called 'plasmoids' (Christiansen *et al.*, 1978) had also been noted by Rees (Rees, 1966).

This all began to change in 1971, when Rees proposed a 'black box' model in which the black box is situated at the centre of certain galaxies, somehow channelling energy into the radio lobes (Rees, 1971). He actually proposed that the energy was transmitted through a tube in the form of a beam of low-frequency radiation. The beam was replaced by a gas jet in 1974 and methods employed in classical fluid dynamics were adapted to the expected environments of galactic cores (Blandford & Rees, 1974). Finally, astronomers were able to detect the jets themselves at radio wavelengths, and some were indeed found to belong to oppositely directed pairs. One of the most stunning early examples was associated with NGC 6251 (Fig. 1.5). The generation of a jet from a blowtorch-like flame within one parsec of the nucleus, yet extending out over 100 000 parsecs without a significant alteration in

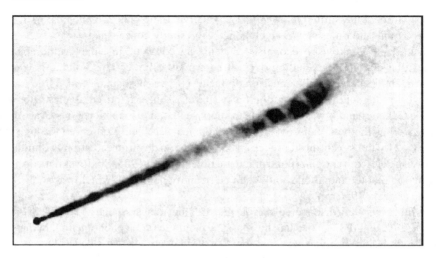

Fig. 1.5 A radio image of the jet associated with the galaxy NGC 6251 taken at 1.4 GHz with the Very Large Array in New Mexico. At a distance of 107 Mpc, the jet extends 140 kpc (280") on this image. The host galaxy is located at the eastern (left) end of the jet. A very faint counterjet has been detected. Credit: NRAO & Werner *et al.* (2001).

direction, is a phenomenal achievement for a natural process. And so, astrophysical jets became an important field of research in astronomy.

The discovery of motions faster than the speed of light within a jet was to radically alter our approach. In 1979, it was reported that features in the 3C 279 jet that were followed over a period of years moved at ten times the speed of light (Whitney *et al.*, 1971), seemingly impossible until one distinguishes between apparent motion and intrinsic motion. Such superluminal motion was subsequently measured in other quasar jets, requiring an explanation to be found that could be lent to the entire class.

1.4   Surprising discoveries

The discovery that young stars could also drive jets followed a similar trail. Again in the 1950s, small 'clots' of optical emission were detected, sometimes aligned or in groups. Being found in dark clouds and regions containing young stars called T Tauri stars, they were immediately associated with the birth of stars (Herbig, 1951; Haro, 1952). The class of clots was soon named Herbig–Haro objects after their initial finders.

At first, the objects were interpreted as protostars and reflection nebulae. However, spectroscopy revealed atomic emission lines which were specifically attributable to shock waves, i.e. regions of impact. What was the catalyst for the impacts? Once more, in the absence of jets, plasmons were invoked, this time in the form of interstellar bullets (Norman & Silk, 1979)

The early 1980s brought new technologies which led towards another explanation. First, bipolar outflows were discovered, often associated with the Herbig–Haro objects (Bally & Lada, 1983). These were wide lobes containing large masses of cold molecular gas moving at supersonic speeds (deduced from submillimetre emission lines produced by molecules of carbon monoxide when excited into rotation). Two lobes were often apparent, straddling a molecular core, with some mirror symmetry somewhat akin to radio galaxy structure. Unlike their cousins, however, the speed of the outflows could be deduced from the molecular lines. It was found that the lobes possessed distinct radial velocities, blue- and red-shifted relative to the central core. This narrowed down the options – the lobes had to be somehow driven by a centrally located, very young star.

Confirmation that jets were present came when, one after another, optical jets, the atomic blowtorches, were revealed on deep CCD images. The first structure, HH 47 in the Gum nebula, was interpreted as a streamer being excited by a jet (Dopita *et al.*, 1982). The optical jets seemed rather inconsequential, carrying low power and feeding Herbig–Haro objects which speed away and disappear into the interstellar medium. Their potential importance to star formation was not fully recognised until the 1990s when jets from protostars were also discovered. The 'most beautiful' twin jets, displayed in Fig. 1.3, were given the appropriately symmetric name HH 212. The HH 212 outflow demonstrates how jets have helped us discover and then locate new protostars.

Meanwhile, in 1980, the discovery of one apparently unique outflow eclipsed all others. A pair of jets were found to emanate from near the centre of a supernova remnant called W 50 about 4.5 pc away, lying in the constellation of Aquila. The jet source was SS 433, an eclipsing binary system with a 13.1 day period. It was the first evolved stellar system (i.e. one which has finished its main period of hydrogen fusion) to display jets. Two oppositely directed jets were found, with material moving ballistically at a speed of $\sim 0.26\,c$, over a quarter of the speed of light. The jets slowly precess over a wide angle as deduced from 'bizarre' moving atomic emission lines (i.e. shifting Doppler shifts) brought to our attention by Margon *et al.* (1979). We now include SS 433 as a member of a class of high-mass X-ray binaries with one component an accreting neutron star and the companion an early-type star undergoing high mass loss. While a few other compact stellar objects in binary systems, including white dwarfs and black holes, have since been added to the jet list, no object has made such an impact as SS 433.

Microquasar is the term now used to describe a particular class of evolved stars which grew in membership in the 1990s (Mirabel & Rodríguez, 1994). These are radio-emitting X-ray binaries, scaled-down versions of quasars with a radio morphology like quasars and high X-ray luminosity. They are powered by black holes or neutron stars which accrete as part of a binary system. The first star-like object to display jets that exhibit relativistic motion was SS 433. Famous objects in this class are Scorpius X-1 (the first known extrasolar X-ray source) and Cygnus X-1 (the first strong black hole candidate). In these cases, the black hole is only a few solar masses and the accretion disc is hot, generating strong X-ray emission. In a number of associated jets, the motion is superluminal. As a result, by following their temporal behaviour over days, we are now able to explore relativistic twin jets in detail (Rodríguez & Mirabel, 1999).

A more recent addition to the jet-nurturing family is an entire class of planetary nebula (PN). As a star of intermediate mass enters the end stage of its life, the red giant or supergiant phase, winds or weakly collimated outflows appear. These so-called bipolar nebulae are not to be confused with bipolar outflows associated with young stars. These structures have long been known and lend planetary nebula their often spectacular mirror symmetry. Their cause is not entirely due to jets, but the interaction of jets with winds turns out to be vital to their structure. A massive, low-velocity wind removes a large fraction of the material during the asymptotic giant branch (AGB) phase and is followed by a higher-velocity, lower-mass wind that snowploughs into the previously ejected material (post-AGB phase). However, in addition to the global structure, fine structure such as knots and jets is found in about half of PN. In this case the knots are called FLIERS (fast low-ionisation emission regions). Although they are seen on opposite sides of the central star along the major axis of elliptical planetary nebulae, there is no substantial evidence that they are driven by jets.

The discovery of bona fide planetary nebula jets goes back to 1985 when Gieseking *et al.* (1985) and Feibelman (1985) found jet-like features (both through imaging and by spectroscopy). The first twin-jet was found to emanate from the star driving the Eskimo nebula (NGC 2392), while the existence of the one-sided jet in Henize 2-36 (He 2-36) has been controversial. Since then, numerous jets have been discovered in the transition stage (proto-planetary nebula) as well as in the PN stage.

Moving into the 1990s, supernovae have provided us with new hosts for jets. The original feature labelled as the Crab Nebula jet (van den Bergh, 1970), however, is non-radial and is probably a fast-moving filament rather than a true jet in the sense discussed here. Still, jets of relativistic particles were suspected to be driven from pulsars, i.e. the rapidly-rotating neutron stars left over from the explosions (Weiler & Sramek, 1988). The ROSAT X-ray telescope finally revealed extended jets from within the Pulsar Wind Nebula associated with the Crab and Vela pulsars. These jets emit synchrotron radiation from relativistic electrons or positrons. Since then, the Chandra X-ray satellite has revealed a few other pulsar jets, suggesting the existence of a new class with fascinating properties.

Gamma-ray bursts (GRBs) are short-lived bursts of gamma-ray photons, the most energetic form of light. Their serendipitous discovery by US military satellites actually goes back to the late 1960s (Klebesadel *et al.*, 1973). However, the cause of the bursts and the central role played by jets were only realised at the turn of the century. Long-duration bursts that last anywhere from two seconds to a few hundreds of seconds are plausibly associated with hypernovae and collapsars in distant galaxies, collapsars being black holes which mark the death of especially massive stars. With the event, a jet of matter at an ultra-relativistic velocity is ejected. The GRB itself and the afterglow at longer wavelengths are from the jet. The anisotropic emission from a beamed relativistic jet remains at a very low level until the Doppler cone of the beam intersects the observer's line of sight, making off-axis GRB jets directly detectable only at long wavelengths and late times (exceeding one year). Convincing evidence that jets are involved was perhaps first presented by analysing GRB 990123, where a predicted break in the light curve was observed. Since GRBs are now detected roughly once per day from random directions of the sky, we have access to a remarkable sample of relativistic jets.

1.5   Overview and points of view

Bringing all astronomical jets under one umbrella is not facilitated by the physics. There are no physical mechanisms or radiation processes to unify a discussion. The flowing material and emitted radiation cover almost every astrophysical possibility. This is because jets stem from a diverse range of objects (Table 1.1) and are launched through a broad range of environments. The variety is no better demonstrated than by comparing their sizes and speeds, as listed in Table 1.2. Their timescales are also simply incomparable. This means that we will need to take a broad sweep of relevant astrophysical radiation and flow processes before we can get familiar with the jet families.

Table 1.1. *The origin. The supplier of the jet flow and the location of the launcher*

Jet type	Source	Supplier	Host/location
Cometary	Vent/Fissure	Ice reservoir	Comet nucleus
Solar spicule	Photosphere	Photosphere	Sun
Solar coronal	Supergranulation	Coronal hole	Sun
Protostellar	Protostar	Molecular core	Molecular clump
Herbig–Haro	Young star	Gas disc	Molecular cloud
T Tauri microjet	T Tauri star	Remnant disc	Interstellar cloud
Planetary nebula	AGB/Post-AGB	Envelope	Interstellar medium
Symbiotic star	White dwarf	Red giant	Binary system
Supersoft source	White dwarf	Star	Binary system
Cataclysmic var.	White dwarf	Red dwarf	Binary system
Low-mass XRB	Neutron star	Low-mass star	Binary system
High-mass XRB	Neutron star/BH	High-mass star	Binary system
Microquasar	Neutron star/BH	Star	Binary system
Pulsar	Neutron star	Pulsar/torus	Supernova remnant
Gamma ray burst	Hypernova/BH	Collapsar	Hypernova remnant
Blazar/quasar	Massive BH	Galaxy nucleus	Massive galaxy
Radio galaxy FRI	Massive BH	Galaxy nucleus	Massive galaxy
Radio galaxy FRII	Massive BH	Galaxy nucleus	Massive galaxy

We will see that jets indeed have a lot in common. There are at least seven themes that recur throughout this work. The unifying themes are related to the dynamics rather than the physics. In addition, the subject also has a natural trisection we are all familiar with: the beginning, the middle and the end, or in jet terminology: the launch, the propagation and the termination.

The first common objective we have is to explain how a directed flow of gas can be produced almost irrespective of the physics. Although a classical description in terms of compressible magnetohydrodynamics may not always be valid, some hybrid plasma version usually is. Secondly, the resulting flow is extended and supersonic. That is, the gas has been accelerated into the channels to speeds exceeding the gas sound speed, and this flow persists for a considerable length. Thirdly, the jets are gradually disrupted or brought to a halt through shock waves, producing concentrated regions of hot, dense gas which pour into gradually evolving reservoirs.

It has often been asserted that there is another significant property common to all classes of jet. This suggested property is related to their near ubiquity: collimated outflows arise wherever gas is in the process of *gravitational accretion*. The idea is that a leaky central engine occurs whenever the two dominant forces involved reach approximate stalemate. The central engine consists of (1) a massive object, providing the dominant gravitational force that sustains a mass supply through gravitational infall, surrounded by (2) a spinning disc, resisting gravity through the outward centrifugal force. While this sets up a centrifugal

Table 1.2. *The scales. Typical sizes, speeds and dynamical timescales which dominate the discussions of each class of astrophysical jet. The speed of light c $\sim$ 300 000 km s^{-1}*

Jet type	Length	Speed	Time scale
Cometary	3 000 km	0.1 km s^{-1}	10 hours
Solar spicule	5 000 km	30 km s^{-1}	10 minutes
Solar coronal	200 000 km	200 km s^{-1}	10 minutes
Protostellar	1.0 pc	100 km s^{-1}	3 000 years
Herbig–Haro	1.0 pc	400 km s^{-1}	1 000 years
T Tauri microjet	0.01 pc	300 km s^{-1}	10 years
Planetary nebula	0.1 pc	300 km s^{-1}	300 years
Symbiotic star	0.01 pc	1 000 km s^{-1}	10 years
Supersoft source	0.01 pc	3 000 km s^{-1}	10 years
Cataclysmic var.	0.001 pc	1 000 km s^{-1}	1 year
Low-mass XRB	3 pc	0.5c	10 years
High-mass XRB	3 pc	0.5c	10 years
Microquasar	0.1 pc	$\sim 1c$	0.1 years
Pulsar	1 pc	0.2c	30 years
Gamma ray burst	0.1 pc	$\sim 1c$	100 days
Blazar/quasar	10 pc	$\sim 1c$	10 years
Radio galaxy FRI	300 000 pc	0.03c	30 Myears
Radio galaxy FRII	300 000 pc	$\sim 1c$	1 Myear

barrier, it is not impenetrable. Energy is dissipated in the disc through frictional forces resulting from the sheared circular motion. Thus material would flow inward if only it could shed its angular momentum. This could be the role of the jets: a channel to extract the waste (excess spin) in order for the remaining gas to feed the central object.

This scenario is, however, hard to prove. Angular momentum could be extracted from the disc in other ways, storing it in planets or transporting it through the disc itself. Thus the connection between the accretion process and the formation of jets remains poorly understood. After examining the facts, we can try to establish it within this book (see section 10.7). If so, then we may better understand how a wide range of astrophysical objects spread throughout our universe have been created.

We are now able to resolve jets as never before. Across the flow, evidence for shear is recovered with lower speeds closer to the periphery than along the axis. In some cases, this may be described as a distinct spine and sheath. We are also attempting to resolve the base of the jet, in both space and velocity where possible. We still need to determine where the collimation is achieved: are jets focused at the point where they are driven (as with comets) or are there distinct driving and collimating mechanisms?

A major aim of jet research is to uncover the driving force. In the first instance, thermal pressure is the obvious candidate to consider. However, there are clear situations where

Table 1.3. *The jet launch. Properties of the driving sources and the probable number of jets per source*

Jet type	Source Size	Detected Disc	Detected Jets
Cometary	1 km	No	Several
Solar spicule	100 km	No	Thousands
Solar coronal	1000 km	No	10/hour
Protostellar	$5\,R_\odot$	Yes	2
Herbig–Haro	$3\,R_\odot$	Yes	2
T Tauri microjet	$2\,R_\odot$	Yes	1/2
Planetary nebula	$100\,R_\odot$	No	2
Symbiotic star	$0.01\,R_\odot$	Yes	2
Supersoft source	$0.01\,R_\odot$	Yes	2
Cataclysmic var.	$0.01\,R_\odot$	Yes	2
Low-mass XRB	10 km	Yes	2
High-mass XRB	10 km	Yes	2
Microquasar	3 km	Yes	1/2
Pulsar	10 km	No	2
Gamma ray burst	10 km	No	1
Blazar/quasar	10 AU	Yes	1
Radio galaxy FRI	1 AU	Yes	2
Radio galaxy FRII	10 AU	Yes	1/2

it is ineffectual and too dissipative, with the predicted radiative flux too high. Radiation pressure is also believed to fall short, barring exceptional circumstances.

We cannot easily find an effective means of transferring gravitational potential into jet energy. When in such doubt, magnetic forces are often invoked even though we have little direct knowledge of their strength or behaviour in the environments of jet sources. However, the 'magnetic models' such as the Blandford & Payne (1982) centrifugally driven model are proving robust. Magnetic fields have the advantage that they can both accelerate winds and collimate them. Strong fields can be induced into differentially rotating discs, which interact with a magnetic field anchored to a central massive object. Combinations of magnetic pressure, magnetic tension and the impulsive acceleration produced when field lines reconnect could provide the mechanism.

The 'sided-ness' is an important qualitative feature (see Table 1.3). Jets are sometimes detected in pairs (e.g. Fig. 1.3). These oppositely directed streams occasionally possess very symmetric structure, implying that variations in the central activity influence the jet and disturbances propagate along both jets at the same speed. We can call these twin jets. Other systems display single jets (e.g. Fig. 5.11). In the case of comets, on the other hand, several jets may appear simultaneously. The number of cometary jets is related to the surface structure, with high-pressure gas venting out through the weakest points on the surface.

In one-sided systems, however, there is usually strong reason to believe that there is intrinsic dynamical symmetry, with twin jets escaping along a specific axis (e.g. the rotation axis of the disc or central object). The evidence is often in the form of bipolar lobes of slow-moving gas or two broad cavities located on opposite sides of the source. These are plausibly interpreted as reservoirs with the jets acting as the pipelines. The most important implication is that there are (or have been) two physical jets, but only one is visible. What could cause such an apparent asymmetry? There is a choice of four excellent explanations, summarised here.

1. *Obscuration.* The disc surrounding the central object is quite stable, although some form of viscosity allows material to fall inwards. However, the process is inefficient and so gas is stockpiled in the disc. Along with the gas, dust particles may remain thoroughly mixed with the aid of some turbulence. The massive disc is thus able to obscure all that lies beyond it. This includes the rear or red-shifted jet unless the jets are aligned with the plane of the sky.
2. *Irradiation.* An external source of ionising radiation shines upon a twin-jet system, exciting and heating the facing jet. It also ablates gas from the surface of a disc, altering the environment which this jet negotiates. On the far side, the second jet remains relatively quiescent.
3. *Beaming.* The gas in the jets moves with a speed close to that of light, c. Relativistic effects concentrate the radiation into two beams of small solid angle. One beam is directed close to the line of sight and is viewed by us from within the solid angle of the beam.
4. *Intrinsic asymmetry.* Alternatively, the two jets may be intrinsically of unequal strength. Due to the nature of the driving source, the gas may be ejected in an alternating pattern (the technical term is flip-flopping). This could be caused by the orbiting motion of the jet source as one component of a binary (black hole or stellar) system. Provided the alternating timescale exceeds the jet dynamical time scale but is exceeded by the outflow lifetime, only one jet exists at any one time.

Besides being one-sided, jets are rarely in any sense uniform. Their brightness varies along and across their lengths, and their shape may be distorted by kinks, pinches and sinusoidal waves. The detected structure is often in the form of compact knots or bow-shaped arcs, suggestive of propagating waves which have steepened into shock fronts within the supersonic flow. By understanding this structure we can learn about how the jet has been formed, what it consists of and how it reacts with the surroundings. To do so, we need to first understand how the radiation is produced (Section 1.6). That is, we need to constrain the physical conditions of the emitting gas.

The second step is to consider the dynamical processes which could lead to the physical state (Section 2.9). Important processes include pulsing (in mass or speed) and wiggling (change in direction) directly from the launch site, wave growth through fluid dynamic forces (e.g. fire-hose instability), the transition flow between laminar and turbulent jets, and magnetic field instabilities.

Jets have important effects as well as causes. The feedback into the environment takes place in the forms of mass, momentum and energy. In particular, momentum and energy transfer are considerable due to the high speed. However, to facilitate the transfer requires the supersonic jet flow to be brought to a halt. The impact regions are called hot spots, terminal shocks or working surfaces, among other terms. They are not generally stationary but continue to push out while compressing, heating and deflecting the gas in the vicinity.

Feedback into the extended environment has even been purported to generate new structures and objects. The compression may trigger episodes of star cluster formation on large galactic scales (in the case of radio galaxies and quasars) as well as of star formation (in the case of protostars).

The impact sites are regions of particularly high pressure. The pressure ensures that the gas spills out. The overflow occupies reservoirs called cavities, lobes and bipolar outflows, as well as being identified with the entire radio galaxy itself. The interface with the environment may produce a dense rim or shell, which may also produce an observable signature.

1.6   Summary

Astrophysical jets are more than just spectacular displays. Some of them provide an effective means of transferring gravitational energy into jet energy, often making an outflow from an inflow. For this reason, we suspect that they are an integral part of the process which leads to the collapse or contraction of mass onto central objects.

However, this is not always the case. Jets in the solar system do not appear to be directly associated with inflow or rotation, and jets from active galactic nuclei may, at least in theory, be the result of energy extraction from a single spinning black hole or a pair of orbiting black holes. In the latter case, it is still the release of spin energy, although not immediately during an object's collapse.

Jets are nevertheless important as signposts: the beacons leading us to objects undergoing tumultuous change. They also provide a history; as with tree rings, past activity can be related to distant features in the jets. Individually, they are also laboratories within which physical processes occur in a concentrated form.

We have reviewed here how the subject has expanded beyond all expectation. Astrophysical jets are almost everywhere. What are the reasons for their occurrence in such diverse objects? If we can put our finger on this, we can learn about the specific from the general. In this regard, there are certain properties of jets which recur in different contexts throughout this book.

We have provided an introduction to seven of the themes which are commonly associated with the jet subject. These themes provide the outline for a systematic comparison. They are: (1) the energy source, (2) the launching mechanism, (3) the collimation process, (4) the jet structure, (5) the penetrating ability, (6) the feedback on the environment, and (7) the destiny of expelled material.

At first sight, angular momentum removal is the obvious cause of the inflow–outflow connection. If there are common dynamical mechanisms, we might infer the nature of central engines simply by comparing the behaviour of their jets. In this respect, jets and accretion discs are often intimately related. More generally, outflow is related to inflow. This view has been promoted because that is what we now want to find: the importance of jets is raised if they prove necessary as spin extractors. With removal of angular momentum from central objects, they can contract or collapse. With removal of angular momentum from discs, material can flow onto the central object.

We have cautioned that the inflow–outflow connection is not exclusive. There is a more general cause for jets which we will attempt to define in this book. The true umbrella term, which objectively covers all the phenomena, is 'supersonic gas dynamics within slender channels'.

2

Detection and measurement

The material contained in the jet is the first property to ascertain. Almost all our information is derived from distant observation. We can only hope to send probes to rendezvous with comets and possibly intercept their jets. Further afield, we rely upon the radiation that reaches us. Therefore, a chapter to outline the radiative properties is a necessity to interface observations with the physical structure.

The evidence is accumulated across the electromagnetic spectrum providing a bank of line strengths and continuum fluxes. These continuum values yield the spectral shape, perhaps a power law with excess emission in the form of broad bumps. The specific wavelengths of the line fluxes can often be confidently identified with the generating atoms or molecules, and the ratios of fluxes may directly yield the particle states. Here, we introduce the major radiation processes to which matter in various states is subject. In the chapters which follow, we may then work backwards to relate the spectral properties to the gas which would produce it.

That, however, is only half the story. We must still question whether or not we are detecting the jet itself. Instead, it may merely be the excited interface with ambient gas or it could be a minor component of the jet. In fact, we often detect just a few discrete regions where the density or temperature has been temporarily enhanced. This can be misleading, since these regions are atypical of the underlying continuous flow which has produced them. This has remained highly problematic for extragalactic jets where we only observe the effects of the relativistic electrons and magnetic field. It may also be a problem for protostellar jets, where cold and neutral atomic and molecular components can be easily hidden unless warming shock waves propagate through the jet. And now, also, the gas jets from cometary nuclei are assumed but not observed to carry the dust away from the surface.

The power carried by a supersonic jet lies mainly in the high-speed bulk motion. The luminosity, on the other hand, depends on the proficiency of mechanisms which tap into this reserve, converting energy of fluid motion into particle motion. The excited particles may then radiate away this energy. Here, we see how to deduce the properties of these particles by understanding the radiative processes. In the following chapter, we will study how the fluid motion may be dissipated into particle energies.

The first step is to establish the type and strength of the radiation. The radiation processes are quite diverse. Synchrotron radiation is produced by particles moving at relativistic speeds and we need to confront the issue of how such speeds are reached. At the other extreme, molecular emission is generated when the constituent atoms rotate or gently vibrate but do not dissociate. We thus pack here the toolbox: the fundamental formulae for a variety

Table 2.1. *The jet emission. An important spectral range and the corresponding radiation processes for each class of jet. Note that these are exemplary; most jet types radiate significantly across several spectral regions. Key: Cont. = continuum; iC = inverse Compton; Rel. = relativistic; Collis./recomb. = collision/recombination*

Jet type	Waveband	Type	Material	Process
Cometary	Optical	Cont.	Dust	Scattering
Solar spicule	Optical	Line	Atomic	Collis./recomb.
Solar coronal	X-ray	Cont.	Thermal	Bremsstrahlung
Protostellar	Infrared	Line	H_2	Rotation/vibration
Herbig–Haro	Optical	Line	Atomic	Collis./recomb.
T Tauri microjet	Optical	Line	Atomic	Collis./recomb.
Planetary nebula	Optical	Line	Atomic	Collis./recomb.
Symbiotic star	Radio	Cont.	Ionised gas	Free free
Supersoft source	Optical	Line	Atomic	Collis./recomb.
Cataclysmic var.	Optical	Line	Atomic	Collis./recomb.
Low-mass XRB	Radio	Cont.	Rel. electrons	Synchrotron
High-mass XRB	Radio	Cont.	Rel. electons	Synchrotron
Microquasar	Radio	Cont.	Rel. electrons	Synchrotron
Pulsar	X-ray	Cont.	Rel. electrons	Synchrotron
Gamma ray burst	γ-ray	Cont.	Rel. electrons	Synchrotron/iC
Blazar/quasar	Radio/optical	Cont.	Rel. electons	Synchrotron
Radio galaxy FRI	Radio	Cont.	Rel. electons	Synchrotron
Radio galaxy FRII	Radio	Cont.	Rel. electons	Synchrotron

of emission mechanisms. We may then proceed to constrain their nature: the speed, mass, energy, and how they are distributed in space and time.

Some basic properties of the radiation are introduced in Table 2.1. Note that the table is not complete – many jets radiate across the entire spectrum and emit both in discrete lines and a broad continuum.

2.1 Synchrotron radiation

The operation of the synchrotron process in radio galaxies led to the first opportunity to establish facts and quantify the properties of extragalactic jets. However, the emission spectra take the form of featureless continua usually approximated with a power-law dependence on wavelength (Ginzburg & Syrovatskii, 1965; Shu, 1991). Despite the lack of information, some useful constraints have been derived for synchrotron jets. To find these, we first consider the emission from a single particle.

The motion of a charged particle in the presence of a magnetic field is a spiral. The circular component around the lines of magnetic field gives rise to centripetal acceleration which prompts the emission of photons. Such radiation is termed *cyclotron* when the particle is not moving with a relativistic speed, *gyro-synchrotron* when speeds are mildly relativistic and *synchrotron* when the speed closely approaches that of light.

The emitted radiated power from a charge q of rest mass m undergoing acceleration a is provided by a simple classical formula: $P = (2q^2/3c^3)a^2$. The non-relativistic centripetal acceleration of $v_\perp^2/r$ is replaced by $\gamma^2 v_\perp^2/r$ in the relativistic case, where $v_\perp$ is its speed transverse to the magnetic field, r is the radius of gyration and the particle Lorentz factor γ is related to the particle speed, v, by

$$\gamma = \frac{1}{\sqrt{1 - v^2/c^2}}. \tag{2.1}$$

For a radio-emitting relativistic electron, we will typically encounter GeV energies where 1 GeV corresponds to $\gamma = 1960$. The force is $ma = dF/d\tau = \gamma d(\gamma mv)/dt = m\gamma^2 v_\perp^2/r$, where $\tau = t/\gamma$ is the proper time and F is the momentum. We also now define β for the particle,

$$\beta = v/c, \tag{2.2}$$

ready for future reference.

The gyroradius is inversely proportion to the magnetic field. It is written in terms of the gyration (or angular) frequency, ω_B, as $r = v_\perp/\omega_B$ which brings in the magnetic field strength, B,

$$\gamma \omega_B = \frac{qB}{mc} = \omega_L, \tag{2.3}$$

where $\nu_L = \omega_L/2\pi$ is called the Larmor frequency. We also introduce the pitch angle, ζ, as the angle between the instantaneous direction of particle motion and the magnetic field: $\sin \zeta = v_\perp/v$. Then, the formula for synchrotron radiation becomes

$$P = \frac{2q^4 B^2}{3m^2 c^5} \gamma^2 v^2 \sin^2 \zeta. \tag{2.4}$$

Note that the dependence on mass and γ implies that we need only consider light particles such as positrons and electrons which can attain highly relativistic speeds. For electrons, we employ the Thomson cross-section $\sigma_T = 8\pi r_e^2/3$ where the classical radius of the electron is $r_e = e^2/m_e c^2$. Then, the power from a single electron can conveniently be expressed as

$$P = \frac{\sigma_T c}{4\pi} \beta^2 \gamma^2 B^2 \sin^2 \zeta. \tag{2.5}$$

Employing the magnetic field energy, $U_B = B^2/(8\pi)$, this is $P = 2\sigma_T c U_B \beta^2 \gamma^2$. To make this formula ready for application, we assume an isotropic distribution of electron velocities to obtain the average power emitted as

$$P = \frac{\sigma_T c}{6\pi} \beta^2 \gamma^2 B^2 = \frac{4\sigma_T c}{3} \beta^2 \gamma^2 U_B. \tag{2.6}$$

The emitting body will now be assumed to be transparent to the radiation it produces. Note that the radiation from an electron is beamed with half within a cone of semi-angle $1/\gamma$ radians. The beam is swept around with the electron motion and so will be observed as a series of pulses, a pulse occurring each time the cone crosses the line of sight. Nevertheless, on assuming isotropy, the received power must average to the emitted power, as given by Eq. (2.5), by energy conservation.

Note also that if the magnetic field is directed close to the line of sight, the pitch angle must be small for the beaming to be in the observer's direction. According to Eq. (2.5), however, the dependence on ζ implies that the radiated power will then be low. We thus tend to select against regions of parallel field in our observations.

Most of the radiation from the received pulses will be boosted to a high frequency $\gamma^3 \omega_B$, due to time dilation (a factor γ) and the cone sweeping or headlight effect (a factor γ^2). The Larmor frequency is $\nu_L = 2.8\,(B/1\,\text{Gauss})$ MHz and the radius of the orbit is $1.5 \times 10^7\,(\gamma/10^3)/(B/1\,\text{Gauss})$ cm. Hence, the characteristic frequency is

$$\nu_c = \gamma^2 \nu_L = 280 \frac{B}{10^{-4}G} \left(\frac{\gamma}{10^3}\right)^2 \text{MHz}. \tag{2.7}$$

The emission from an ensemble of particles is governed by several parameters. From a single electron, however, the spectrum generated is actually discrete, but the frequency spacing between the harmonics at which an electron radiates is very narrow. In any case, a continuous spectrum will always be observed simply because a vast number of relativistic particles with a wide range in energies are invariably encountered. Empirically, the energy distribution is approximated by a power-law function

$$n(\gamma) = n_o \gamma^p, \tag{2.8}$$

where n is the number density per unit γ. This distribution determines the observed spectrum provided the plasma is optically thin.

We take the radiation of an electron to be received at the frequency $\nu = \gamma^2 \nu_L$ and an isotropic distribution for the pitch angles. Integration over the energy spectrum yields an emissivity

$$J_\nu \sim \frac{c \sigma_T B^2}{12 \pi \nu_L} \left[\frac{\nu}{\nu_L}\right]^\alpha, \tag{2.9}$$

where the *spectral index* is

$$\alpha = (1+p)/2. \tag{2.10}$$

We have taken the limit of high γ, which is usually justified for extragalactic jets. The spectral index can be seen to derive from the fact that $\nu \propto \gamma^2$, which gives $d\gamma \propto \nu^{-1/2}\,d\nu$.

The power-law index, p, is attributed to the relativistic electron acceleration process. Observed jet spectral indices are in the range $\alpha \sim [-0.5, -1]$. Therefore, the range in p is $\sim [-2.0, -3.0]$.

The total energy in relativistic electrons can be constrained by integrating Eq. (2.8) across all energies. Given a steep energy distribution with $p < -2$, most of the energy lies near a presumed low-energy cut-off, γ_{min}. This can be related to the lowest emitted frequency that is observed in order to derive a minimum cosmic-ray electron energy, U_{CR}. For a given observed radio source, one can then minimise the total energy, $U_B + U_{CR}$, required to reproduce the synchrotron spectrum. This yields

$$U_{CR} = \frac{4}{1-p} U_B, \tag{2.11}$$

which happens to be extremely close to what is called the *equipartition requirement*, $U_{CR} = U_B$, for the range in p noted above (Demoulin & Burbidge, 1968; Shu, 1991). However, there is no a-priori reason to believe that the magnetic fields and relativistic electrons in radio sources are in equipartition. Either of these condition may make some physical sense if there is an efficient mechanism to transport energy between the two forms which favours the appropriate direction. However, there is now growing evidence that, at least for radio galaxies, these energies may be small in comparison with that carried by the jet and deposited into the lobes in the form of relativistic protons (see Chapter 11).

The timescale for the depletion of the relativistic energy places a second constraint on models. The synchrotron loss timescale, t_s, is derived from Eq. (2.6) on equating $P = -(d\gamma/dt)m_e c^2$ to yield

$$t_s = \frac{-\gamma}{d\gamma/dt} = \frac{3m_e c}{4\sigma_T U_B \gamma}. \tag{2.12}$$

This implies that the highest-energy relativistic electrons are the first to lose their energy in the absence of re-acceleration. This energy is related to the frequency as given by Eq. (2.7).

We thus expect the power-law spectrum to be cut off at a high-frequency break or turnover. Without re-acceleration, the break energy moves with time to lower values, and a knee in the spectrum at frequency ν_{br} also shifts to lower frequencies as given by

$$t_s = 10^7 \left(\frac{B}{10^{-4}\,\mathrm{G}}\right)^{-3/2} \left(\frac{\nu_{br}}{1\,\mathrm{GHz}}\right)^{-1/2} \mathrm{year}. \tag{2.13}$$

As will be seen, this timescale is not too short for radio jets to remain detectable even over hundreds of kiloparsecs. However, it places constraints on the ages of radio lobes, which must be replenished on this timescale. Moreover, if the relativistic electrons which produce optical and X-ray jets are only accelerated at the base, within a galactic nucleus, then with $\gamma \sim 10^6$ their lifetime becomes shorter than the propagation time.

2.2   Self-absorption and polarisation

We have assumed that the plasma in which the jet propagates is optically thin to its own synchrotron radiation. We can show that a very compact synchrotron-emitting source will instead be optically thick for sufficiently low frequency radiation propagating through the enclosed relativistic electrons. More precisely, the opacity for synchrotron self-absorption is $\kappa \propto \nu^{(p/2-2)}$. If it is optically thick, then black-body physics yields the intensity as the Planck function:

$$I(\nu)d\nu = \frac{2h\nu^3}{c^2} \frac{1}{\exp\left(\dfrac{h\nu}{kT}\right) - 1} d\nu, \tag{2.14}$$

for the energy radiated per unit area and per unit time in the frequency interval ν to $\nu + d\nu$. We take the Rayleigh–Jeans (low frequency) limit, $h\nu \ll kT$ and substitute the electron energy $\gamma m_e c^2$ for the particle energy kT where, from Eq. (2.7), $\gamma \propto B^{-1/2}\nu^{1/2}$. This yields a rapidly increasing intensity for the form of the low-frequency spectrum as $I \propto B^{-1/2}\nu^{5/2}$. Note that this does not depend on the relativistic electron energy distribution, i.e. the index p.

Such steep 'inverted' spectra are *not* often encountered in the context of jet synchrotron sources. Instead, flat or undulating spectra are found and several explanations have been offered for this difference. First, at the base of a diverging jet flow, where progressively higher frequencies sample progressively deeper into the throat, the resultant spectrum may conspire to be quite flat (Reynolds, 1982). In this case, the shape of the spectrum is independent of the energy distribution of the relativistic electrons (i.e. the index p above). This is the classical interpretation of compact flat-spectrum sources and contrasts with the optically thin synchrotron spectra associated with extended steep-spectrum sources such as radio galaxies.

A second explanation is in terms of a superposition of a quite large collection of emitting regions each possessing a core which becomes optically thick below a different frequency. When added together, a flat undulating spectrum can be generated. Finally, a flat spectrum can result from an optically thin region provided the distributions of the magnetic field and relativistic electrons in space both conspire in a precise manner.

An independent constraint on the magnetic field strength can be found if we find the frequency at which the optically thin synchrotron spectrum reaches a peak. (Fluxes and source sizes at lower frequencies become difficult to measure.) The observed flux, F, from an optically thick source is proportional to the intensity, I, and the solid angle, Ω, subtended by the source. We thus obtain $F \propto B^{-1/2}\nu^{5/2}\Omega$. This formula provides a constraint on the magnetic field strength on measuring the flux and frequency of the spectral peak, provided we know the angular size.

A high polarisation of the synchrotron radiation will be produced from an optically thin blob pervaded by a uniform magnetic field. The intrinsic polarisation is perpendicular to the projected magnetic field direction in this optically thin case. A detailed discussion requires an analysis of the Stokes parameters. Here, we note the relevant result that the degree of polarisation as defined by

$$\pi_L = \frac{P_\perp - P_\parallel}{P_\perp + P_\parallel}, \tag{2.15}$$

where the power is split into components parallel and perpendicular to the field projected onto the sky, yields

$$\pi_L = \frac{1-\alpha}{5/3-\alpha} = \frac{1-p}{7/3-p}, \tag{2.16}$$

which is 69% for $p = -2$ and 75% for $p = 3$.

Given N cells of magnetic field each with a uniform field with direction selected from a random distribution, the resulting polarisation would be reduced by a fraction $\pi_p/N^{1/2}$. Furthermore, in the optically thick case, the maximum polarisation is reduced to about 10%, with the direction of polarisation now parallel to the magnetic field. However, a disordered but anisotropic magnetic field distribution can also produce a high polarisation. Either shear or compression along a single direction, such as produced at a jet boundary layer or by a planar shock wave within a jet, will create a preferred direction that may result in a high degree of polarisation (e.g. Hogbom, 1979; Laing *et al.*, 2006).

Circular polarisation from incoherent synchrotron radiation is expected to be very low, being reduced by the mean electron Lorentz factor which corresponds to the emitted frequency as given by Eq. (2.7) (remembering that the observed critical frequency is increased by the Doppler factor).

The polarised radiation emitted by a given set of electrons is affected by Faraday fog: the process of rotation of the intrinsic polarisation angle through other screens of electrons and magnetic field along the line of sight. A linearly polarised wave can be decomposed into left-handed and right-handed circularly polarised waves which propagate at different phase velocities, resulting in a rotation of the electric field vector. The total rotation in the angle of polarisation depends on the strength and orientation of the intervening magnetic field, the electron density and the wavelength, λ. Passing through a slab of thickness L containing a uniform field, the rotation angle is given by RM $\times \lambda^2$, where the rotation measure RM $\propto n_e B_\parallel L$ with n_e the electron density and $B_\parallel$ the field component along the line of sight. If internal to the emitting plasma, the amount of rotation will depend on the Faraday depth, thus leading to depolarisation which can be attributed to Faraday rotation if the specific wavelength dependence is found (Zavala & Taylor, 2004). It is critical to first distinguish foreground from internal Faraday rotation. Much of the rotation may take place outside the source and there may also be many field reversals along the line of sight, severely complicating an interpretation.

2.3   Compton processes

The same relativistic electrons which produce the photons will also scatter them. The upscattering in energy is called the inverse Compton process, which is equivalent to Thomson scattering in the electron's rest frame. Photon energies are increased by a factor $\sim \gamma^2$ upon scattering off electrons of energy $\gamma m_e c^2$ in the observer's frame. Hence, inverse Compton scattering cools the electrons by a mean factor

$$\frac{4}{3}\beta^2\gamma^2 U_{rad} c\sigma_T, \tag{2.17}$$

which compares with the synchrotron cooling with the photon energy density U_{rad} in place of the magnetic energy $U_B = B^2/(8\pi)$. Thus, the ratio of Compton to synchrotron losses is just U_{rad}/U_B.

The above result has important implications for quasars where extremely high radiation energy densities are deduced. Assuming that the radio emission is synchrotron radiation, the predicted inverse Compton losses should dominate, and higher-order scatterings would lead to a runaway process – the Compton catastrophe – discussed in Section 5.4. This would, however, in reality be limited by the finite energy resource of the electrons (mathematically, the Thomson cross-section is replaced by the Klein–Nishina cross-section).

The brightness temperature is often the starting point for interpreting radio quasars and other flat-spectrum sources. This has meaning in the optically thick part of the spectrum where, from Eq. (2.14), we define

$$T_b = \frac{c^2 I}{2k\nu^2}, \tag{2.18}$$

which corresponds to the low-frequency end of the black-body spectrum. This is frequency-dependent but peaks at a frequency ν_m, which separates the self-absorbed spectrum $\propto \nu^{5/2}$ from the optically thin synchrotron $\propto \nu^{-\alpha}$.

Measured values T_b were found to reach extreme values, exceeding 10^{11} K, which imply a relativistic plasma (Kellermann & Pauliny-Toth, 1969). However, values exceeding 10^{12} K have often been found using the variability timescale, t_{var}, to limit the source size of spatially unresolved radio sources to ct_{var}. This is indeed a problem, since inverse Compton losses should stringently limit the brightness temperature to 10^{12} K (Kellermann & Pauliny-Toth, 1969). It is now believed that these radio cores correspond to the synchrotron self-absorption bases of conically expanding, highly relativistic jets in which beaming effects greatly boost the observed intensities.

2.4 Electrons: free-free and bremsstrahlung processes

Thermal bremsstrahlung or free-free emission is produced during the Coulomb interaction of charged particles. The power generated from encounters of all electrons with density n_e with ions of charge Z_i and density n_i can be derived from classical approximation,

$$P = n_e n_i \left[\frac{2m_e}{3\pi kT} \right]^{1/2} \left[\frac{32\pi^2 Z_i^2 e^6}{3m_e^2 c^3} \right] g_{ff} e^{\frac{-h\nu}{kT}}$$

$$= 6.8 \times 10^{-38} n_e n_i Z_i^2 T^{-1/2} g_{ff} e^{\frac{-h\nu}{kT}} \text{ erg s}^{-1} \text{ cm}^{-3},$$

(2.19)

on assuming a Maxwellian velocity distribution and with g_{ff}, the Gaunt factor, a slowly varying function of ν that accounts for quantum mechanics.

Note that a flat spectrum is generated at low frequencies, $h\nu \ll kT$, from a gas at constant temperature. The radiation is also distinguished by a lack of intrinsic polarisation. Integrating over all frequencies, we obtain the power radiated through thermal bremsstrahlung:

$$P_t \sim n_e n_i \left[\frac{2m_e}{3\pi} \right]^{1/2} \left[\frac{32\pi^2 Z_i^2 e^6}{3m_e^2 c^3 h} \right] g_B (kT)^{1/2},$$

(2.20)

where $g_B(T)$ is a frequency-averaged Gaunt factor. Thus, given that $n_e = n_i$, free-free emission is a sensitive probe of the electron density.

The low-frequency part of the spectrum will be optically thick, with a black-body like dependence of frequency, $P \propto \nu^2$. At high frequencies, the Gaunt factor provides a shallow decrease of the form $P \propto \nu^{-0.1}$.

Radio emission is proving to be a powerful means of investigating the sources of jet phenomena associated with forming stars (Section 6.5). In some cases, the source of the emission is resolvable into small-scale jets within 100 AU. In unresolved cases, we may expect the spectral shape to be determined by the expansion properties of the jet, which will consist of an opaque base and an extended transparent zone. For these partially opaque jets, we predict spectra with indices anywhere between the optically thick and thin limits, consistent with the observations (Reynolds, 1986). In other words, it is the dynamics and geometry, as given by the degree of expansion and acceleration, that appear to determine the radio spectrum.

2.5 Atomic processes

The detection and identification of line emission established the presence of atoms and molecules in certain classes of jets, as amply recorded in Table 2.1. The emission lines are the primary means of probing the physical conditions in non-relativistic stellar jets. The line wavelengths in the optical correspond to those of atoms of hydrogen and helium, as well as many trace elements such as sulphur, nitrogen and oxygen in partially ionised forms.

Specific sets of lines allow a direct and detailed diagnosis of a jet, in stark contrast to the limited information derivable from synchrotron jet spectra. The variation of electron and atomic density along the jet, as well as the temperature, can be deduced from the relative fluxes of pairs of lines (line ratios). This follows when the atoms are collisionally excited or ionised, followed by collisional or radiative de-excitation with rates that depend on the number of collision partners and the impact speeds. Even more directly, each line profile shape, line shift and line width provides independent information on the source dynamics. The line shift is the difference between the measured wavelength and the laboratory wavelength of the line in vacuum, caused by the motion of the emitting material along the line of sight.

The spectrum corresponding to the hydrogen atom is dominated by emission from recombination cascades. Following recombination of an electron with an ion, usually into a state with high principal quantum number, n, the electron cascades down towards the ground state, producing emission in resonance lines along the way. Transitions into the ground state $n = 1$ are called Lyman transitions and those into the first excited state $n = 2$ are called Balmer lines. In stellar jets, the Hα line at 6563 Å and Hβ at 4861 Å, the transitions from $n = 3$ and $n = 4$ to $n = 2$, respectively, have been the starting point of the majority of studies.

Strong 'forbidden' lines are produced as well as the permitted lines. These lines are suppressed in high-density gas, being forbidden by dipole selection rules. However, they are permitted by quadrupole selection rules, albeit with very low transition probabilities. At high density, such as encountered in the laboratory, the rate of collisional excitation and de-excitation will exceed the rate of spontaneous emission and the line would not be seen. In contrast, in the near vacuum conditions of the interstellar medium, collisions are infrequent and the low probabilities for radiative transitions are still sufficient to produce emission lines. Furthermore, the low transition probability implies that the photons will generally not be re-absorbed and so will escape the region from which they originate.

The density related to a jet can be constrained in a number of ways. For example, the need to supply sufficient momentum and kinetic energy into the extended source constrains the average density. Independently, the emitted lines provide a measure of the jet density from within the emitting regions. To each energy level there corresponds a transition range of densities above which the level is depopulated according to collisions and below which the level is depopulated by spontaneous emission. This is characterised by the critical density at which the two means of escape have equal probabilities. Hence, given two lines with different critical densities, their ratio provides a constraint on the density (of the collision partner) provided the density lies near the transition range. Note also that at low densities the line flux (emissivity) is proportional to the density squared while the flux is linearly dependent on density above the transition range.

Jets expand laterally over many orders of magnitude and the density is expected to fall rapidly. It is thus realised that we need multiple diagnostics and high spatial resolution, the basis of spectro-astrometry (Whelan *et al.*, 2004). One common diagnostic is the [S II] 6716 Å/6731 Å flux ratio, R, which is approximated by $\log(R) \sim \log(n_e/10^3\,\mathrm{cm}^{-3})$ in the transition range $n_e = [2 \times 10^2 \, \text{cm}^{-3}, 1 \times 10^4 \, \text{cm}^{-3}]$ for the electron density (taking an electron temperature of 10^4 K). At the base of young stellar object jets, the [Fe II] 7155 Å/8617 Å ratio is more appropriate since it measures the electron density in the denser regions, with $R \sim \log(n_e / 10^6 \, \text{cm}^{-3})$.

The infrared also contains atomic lines which are detected from jets, but because of lack of sufficient sensitivity and spatial resolution, they have not been much utilised until recently. Among these are the hydrogen Paschen β (1.2822 μm) line and several [Fe II] lines (Whelan *et al.*, 2004).

2.6   Molecular processes

Molecules are associated with jets in planetary nebula and young stars. The major detections to date have corresponded to spontaneous line emission from the diatomic molecules CO, SiO and H_2. It is also clear that the jets are likely to contain many other molecules as well, and they may turn out to be vital tracers. In addition, through shock heating and compression during the interaction with ambient gas, the composition is modified and more complex molecules are also formed and excited. Their subsequent emission provides further constraints on the properties of the driving jet.

Molecular emission falls into three categories, which involve rotational, vibrational and electronic transitions. The lines or bands tend to fall in the millimetre/submillimetre, infrared and optical/ultraviolet, respectively, due to the typical energies involved. First consider a diatomic molecule which is not electronically or vibrationally excited. Diatomic molecules in rotational motion possess quantised energy levels, E_J, which can be written in the form

$$E_J = J(J+1)\frac{h^2}{8\pi^2 I}, \tag{2.21}$$

where J is the rotational quantum number and I is the moment of inertia of the molecule.

Carbon monoxide is a common molecule found in protostellar jets and associated outflows. For $^{12}C^{16}O$, substituting values for I, we obtain the rotational constant $B = E_1 - E_0 = 2.77$ K. Transitions are limited by the selection rule $\Delta J = 0 \, or \pm 1$. Therefore, radiative transitions are limited to $\Delta J = 1$ and the emitted photons carry away one unit $(h/2\pi)$ of angular momentum. The difference between energy levels is given by

$$\Delta E_J = (J+1)\frac{h^2}{4\pi^2 I}, \tag{2.22}$$

which yields emission in the millimetre and submillimetre ranges at frequencies 115 GHz (2.6 mm), 230 Ghz (1.3 mm), 346 GHz etc. which is the sequence of R-branch transitions (J decreases by 1). The $J = 1$ level tends to be well populated given a modest radiative rate of $A_{10} = 6 \times 10^{-8} \, s^{-1}$. This line, moreover, can be optically thick, which also tends to raise the $J = 1$ level population. Hence, emission arising from low-J energy levels is often detected from gas of temperature 10 K to 60 K, tracing cold carbon monoxide gas in the jet.

The production of silicon monoxide within jets is mainly due to the reprocessing of dust within discrete shocked regions called molecular bullets. Shocks with speeds exceeding $\sim 20 \, \text{km s}^{-1}$ release silicon from grains into the gas phase, followed by rapid formation of SiO through reactions involving the OH molecule. Hence, pure rotational emission from high-J SiO including $J = 11-10$ and $J = 8-7$ at 478 GHz and 347 GHz, respectively, have been detected along the HH 211 protostellar jet (Nisini *et al.*, 2002), in addition to the SiO $J = 1-0$ line at 43 GHz, $J = 3-2$ line at 130 GHz and the $J = 5-4$ line at 217 GHz. This demonstrates that jets may contain warm (>250 K) and dense $n(H_2) \sim 5 \times 10^6$ cm^{-3} molecular gas.

The hydrogen molecule has a much lower moment of inertia, which results in infrared emission from warm gas. However, it is homonuclear with no permanent dipole moment. The exchange symmetry associated with the two identical atoms permits only quadrupole transitions of the form $\Delta J = 0, \pm 2$. The nuclear spins ($\pm 1/2$) may be parallel ($S = 1$) or anti-parallel ($S = 0$) with statistical weight $2S + 1$. To maintain an overall anti-symmetry, molecular hydrogen exists in two forms or modifications. Molecules with odd J-values and $S = 0$ are called para H_2 while those with even J-values and $S = 1$ are called ortho H_2. Nevertheless, a molecule can still be modified after formation given sufficient time and the appropriate types of collision.

The most common tracer of molecular hydrogen in jets is the infrared 2.12 μm line which lies within the K-band, observable with ground-based telescopes. This corresponds to the 1–0 S(1) transition, from the first to the ground vibrational level, while the rotational S-transition corresponds to J decreasing by 2 into the $J = 1$ level. The upper level of the 1–0 S(1) transition corresponds to a temperature of 6956 K. The widespread detection of this line from within jets thus testifies to shock excitation with shock speeds exceeding ~ 10 km s^{-1} being capable of producing copious gas with temperatures of a few thousand kelvin. There are a large number of other ro-vibrational transitions which have been detected from the jets themselves, including emission stemming from higher vibrational levels combined with O-branch (J constant) and Q-branch (J jumps up by 2) transitions.

For both CO and H_2, the line fluxes can be directly related to the mass and excitation temperature, T, of the gas on assuming local thermodynamic equilibrium (LTE). The emitted flux, F_{ij}, of a line at frequency ν_{ij} is related to the column of gas in the upper energy level of the transition, N_i, within the observing beam by $F_{ij} = N_i A_{ij} h \nu_{ij}$ where A_{ij} is the Einstein coefficient (spontaneous transition probability). If in LTE, $N_j = g_j Z N_0 \exp(-T_j/T)$ where kT_j is the energy of the upper level, g_j is the level degeneracy and Z is the partition function. For rotationally excited CO this yields a density distribution $n_J = n_0(2J+1)\exp(-J(J+1)B/kT))$, with LTE appropriate for densities $n_0 > 10^4$ cm^{-3}.

For molecular hydrogen, modelling of the observed near-infrared emission fluxes indicate excitation temperatures in the range [1400 K, 3000 K]. However, LTE would only be approximately reached for densities exceeding 10^7 cm^{-3} in a pure H_2 gas or 10^5 cm^{-3} if hydrogen atoms are also present in sufficient quantities (i.e. $n(H) > 0.01 n(H_2)$ since an atom can generally penetrate deeper into the molecule during an encounter.

2.7   Maser beams

Maser emission is possible when the populations of two energy levels become inverted (Elitzur, 1992). To sufficiently enhance the population in the upper energy level (level 2) requires very particular circumstances. We require a mechanism to efficiently pump an even higher level (level 3). Then, a rapid cascade down into the upper level (level 2) can produce the required inversion. The pumping mechanism can be based on collisional or radiative excitation. When masering is effective, a photon with energy corresponding to the transition energy, $h\nu_{12}$, stimulates the production of additional photons moving in the same direction, since the absorption is negative. The emission is highly beamed since the negative absorption requires low velocity differences to operate. It follows that the emission is amplified rather than being attenuated along the beam.

The maser gain, the exponential growth with distance, is limited by the rate at which the molecule is pumped into the highest level, which is itself limited to the spontaneous emission rate, A_{13}. Given a maser beam of solid angle Ω and a maser transition of frequency ν_{12} and spontaneous emission rate A_{12}, this yields a maximum brightness temperature

$$T_B \sim \frac{A_{13}}{A_{12}} \frac{4\pi}{\Omega} \frac{h\nu_{12}}{k}. \tag{2.23}$$

Hence, even with a thermal temperature of 100 K (effectively the third factor), beaming of $\Omega = 0.001$ and $A_{13} = A_{12}$ yields T_B of 10^{12} K.

Maser spots are thus detected, representing those locations where the narrow cone of stimulated emission is directed towards us. In the context of jets, H_2O, OH and SiO masers are observed, although in many cases the radiating material may be the swept-up circumstellar envelope. Other inversions occur in methanol (CH_3OH), ammonia (NH_3) and formaldehyde (H_2CO), which also prove useful tracers of conditions and structure. In particular, Class I methanol masers appear to be directly related to outflow phenomena.

2.8   Power and size

To determine the total luminosity for any object, we require a complete spectrum and the distance. This has been a problem for extragalactic sources. Fortunately, we now have confidence in the distance as determined by the redshift, z, and the Hubble parameter, H, as constrained by the WMAP satellite and other techniques. Where necessary, we assume a standard flat Λ-dominated cold dark matter (CDM) universe, with $H = 72$ km s^{-1} Mpc^{-1} and other parameters defined by Spergel *et al.* (2003).

Remarkably, the distances to the jets in our own galaxy are now less well known, since their driving stars do not participate in the systematic expansion. The Hipparcos satellite has helped placed some constraints, but most of the objects are too distant for parallax or proper motions to as yet provide useful limits.

For a continuum source of radiation, such as a synchrotron jet or protostellar core, we record the total flux density at specific frequencies. The flux densities, F, of many radio galaxies are conveniently expressed in units of janskys (Jy), where 1 Jy is 10^{-23} erg s^{-1} cm^{-2} Hz^{-1}.

A source radio power can be derived from this flux density, although the distance is quite a complex function of the redshift and the deceleration parameter. At low redshift, the spectral luminosity can be written in terms of distance, D, as

$$P_\nu = 1.2 \times 10^{24} \frac{F_\nu}{1 \text{ Jy}} \left[\frac{D}{100 \text{ Mpc}} \right]^2 \text{ W Hz}^{-1}, \tag{2.24}$$

which has assumed a source to be isotropic. In general, a so-called K-correction which depends on the spectral index is also required to transform the radio power to that predicted at the emitted frequency rather than the observed frequency.

At low redshift, a source of size L at a distance D subtends an angle

$$\theta = 0.1 \left(\frac{L}{1\,\text{pc}} \right) \left(\frac{D}{1\,\text{Mpc}} \right)^{-1} \text{arcseconds.} \tag{2.25}$$

Therefore, a parsec-scale extragalactic jet at 100 Mpc subtends an angle of order of a milliarcsecond. In contrast, a 100 kiloparsec scale radio galaxy would stretch across arcminutes. In galactic terms, a jet at a distance of 1 kiloparsec could be resolved down to a few AU with a radio telescope with milliarcsecond resolution, while a parsec-scale protostellar flow would again cover arcminutes.

We often desire to know in which part of the spectrum most energy is emitted. The total luminosity (bolometric luminosity) can be estimated when the jet is emitting with a power-law spectrum over a known range of frequencies. More generally, however, a complete spectrum from low radio frequencies to gamma rays will contain several bumps and breaks and it is not clear what frequency range dominates. To evaluate such a spectrum, it is more useful to plot the spectral *energy* distribution, the SED. That is, we plot vP_v as a function of frequency rather than the flux distribution. The location of the broad peaks on such a diagram can automatically (but roughly) be attributed to the spectral regimes with the highest luminosity, without recourse to integration.

2.9   Summary

As we now record as much detail as we can across the entire spectrum, from radio to gamma rays, the background information provided above should help indicate the underlying physical state of a jet. This will constrain the nature of the emitting material.

We have therefore investigated here what the jets must contain in order to explain directly what we observe. However, this does not explain how the matter reached the observed condition. Not just the observed location, but also the agitation must be accounted for. Identification of the radiation process does not reveal the dynamical mechanism which has accelerated the particles or amplified the magnetic field. Some open questions are as follows. How are particles accelerated to ultra-relativistic speeds? Where do the molecules form? What heats the atoms and molecules?

3

The dynamical toolbox

Jets can now often be resolved. However, the structure revealed is only the distribution of the radiation received along and across a jet. The physical structure may be very different, with shock waves, relativistic beaming and projection influencing how we perceive the flow pattern. Therefore, we need to compare the images with those predicted by theories which incorporate the radiation physics, flow physics and geometry.

Connecting equations are required to relate the structure of a jet to the dynamics and physics. For example, the collimation of an observed jet is described by the angle at which it opens up. Hydrodynamic principles dictate that this angle is related to the Mach number. Here we establish these principles as working tools with which we can handle the various sets of astronomical phenomena that we encounter in the subsequent chapters.

We must also determine the entity that is actually detected. The entire jet may not be visible. We may see just a narrow spine, an outer skin or embedded clumps. Perhaps we only see the outer sheath of entrained ambient material.

Here, we look at the range of fundamental dynamics pertaining to jet flows within scenarios which find some support. We introduce useful parameters such as the Mach number, Alfvén number and Reynolds number. Finally, we discuss shock waves, which are the inevitable consequence of supersonic flow.

3.1   The inviscid hydrodynamic equations

Certain jet models require little more than the equations of fluid dynamics. First, consider the equations for non-relativistic hydrodynamics. The continuity equation expresses the conservation of mass as

$$\frac{\partial \rho}{\partial t} + \nabla \cdot (\rho \mathbf{v}) = 0 \tag{3.1}$$

in terms of the density ρ and velocity vector $\mathbf{v}$.

The physical state requires two further parameters to be specified: the pressure, p, caused by thermal motions and the temperature, T. These are related by an equation of state, and a perfect gas law normally suffices for jet flows:

$$p = nkT = \frac{\rho kT}{\mu m_H}, \tag{3.2}$$

where n is the particle number density and μ is the mean mass of a particle in units of the mass of the hydrogen atom, m_H. However, more generally, the pressure is not isotropic and cannot be treated as a scalar. It needs to be considered as a tensor quantity which then introduces the viscosity through the stress tensor (see below).

The Euler equation expresses how the gas is accelerated by the gradient in pressure and any other external forces, **F**, such as gravity:

$$\rho \frac{d\mathbf{v}}{dt} \equiv \rho \frac{\partial \mathbf{v}}{\partial t} + \rho \mathbf{v} \cdot \nabla \mathbf{v} = -\nabla p + \mathbf{F}. \tag{3.3}$$

The total gravitational acceleration on unit mass can rarely be expressed as a constant **g** and often as the gradient of a potential, i.e. $\mathbf{F} = \rho \nabla \Phi$. This is actually three equations in three directions. The right-hand side is the force on a fluid element and the left-hand side is the response of a fluid element, split into terms corresponding to change with time and space (advection).

Finally, the energy equation takes into account the complete energy budget. It balances the following contributions: (1) the energy contained in the bulk motion of a fluid element; (2) the energy per unit mass stored in internal degrees of freedom of the particles, ϵ; (3) the work done by an external force; and (4) other gains and losses from the element. We thus have

$$\frac{\partial}{\partial t} \left(\frac{\rho v^2}{2} + \rho \epsilon \right) + \nabla \cdot \left[\left(\frac{\rho v^2}{2} + \rho \epsilon + p \right) \mathbf{v} \right] = \mathbf{F} \cdot \mathbf{v} - S_{rad} - S_{con}. \tag{3.4}$$

Here, the flux terms S_{con} and S_{rad} account for the total loss of energy through conduction and radiation (including sources as well as sinks).

The internal energy is related to the pressure by $\epsilon = p/(\gamma - 1)$ according to equipartition between degrees of freedom. The specific heat ratio, γ, is defined as $\gamma = c_p/c_v$, which is the ratio of the specific heat at constant pressure, c_p, to the specific heat at constant volume, c_v.

A simpler form of the energy equation is found by considering just the terms which directly influence the internal energy. This is found by subtracting off the terms related to change of momentum along the direction of motion to yield

$$\rho \left(\frac{\partial \epsilon}{\partial t} + (\mathbf{v} \cdot \nabla) \epsilon \right) = -p(\nabla \cdot \mathbf{v}) - S_{rad} - S_{con}. \tag{3.5}$$

Thus we have six unknowns and six equations. We can add extra terms and equations as we introduce more detailed physical processes including those associated with radiation, conduction, magnetic fields and gravity. While supersonic jets possess speeds far exceeding the gravitational escape speed, they must be launched from sites where gravity will be a constraining force. We can probably ignore self-gravity as far as the jet itself is concerned and only consider external fixed masses, a fact which brings considerable simplification.

Inviscid flow is the simplest approach to jet modelling. However, jets generate shear between themselves and the ambient medium. In addition, the jet itself is likely to be sheared when launched, and so prone to shearing instabilities. In fluid dynamics, shear is usually found to introduce strong viscous effects which lead to turbulence. If the region of strong shear is confined to a narrow interface or boundary layer, the flow in the body of the jet may remain inviscid. The boundary layer, whether turbulent or sheared, can then be represented by a tangential discontinuity in velocity. Hence, provided viscous flow is constrained to occupy narrow surface layers, the inviscid approximation is justified.

3.2 Viscosity

The physical regimes in which an inviscid flow description holds, and in which viscous effects are confined to negligible boundary layers to the jets, are calculated here. The effect of viscosity is included in the Navier–Stokes equation,

$$\rho \frac{d\mathbf{v}}{dt} = -\nabla p_t + +\rho v_m \Delta v, \tag{3.6}$$

where the kinematic viscosity, v_m, is proportional to the the root mean square particle speed, v_{th}, and the mean free path between particle collisions, λ_p:

$$v_m = \lambda_p v_{th}/2. \tag{3.7}$$

For simplicity we avoid the introduction of a stress tensor and instead here use the symbol Δv as the *diffusion* operator, which signifies that the viscous force on a fluid element depends on the change in gradient with distance.

The dominance of viscosity on small scales is shown by a dimensional analysis of the terms in Eq. (3.6). Velocity changes across a jet of order U over a length scale L are presumed. Thus the velocity gradient is of order U/L and the dynamical time is of order L/U. Advective acceleration is of order U^2/L, whereas viscous acceleration is of order $v_m U/L^2$. Hence, viscosity is most effective on small scales. The ratio of these advective and viscous acceleration terms defines the fluid dynamic Reynolds number.

For a neutral gas of density n, elastic collisions yield

$$\lambda_p = \frac{1}{\sigma_p n}, \tag{3.8}$$

where σ_p is the collision cross-section, of order of 10^{-15} cm^2 for neutral species. The sound speed in an isothermal molecular jet can be written in the form

$$c_s = 0.44 \mu_m^{-1} \left(\frac{T}{10\,\mathrm{K}} \right)^{1/2} \mathrm{km\ s}^{-1} = 0.19 \left(\frac{T}{10\,\mathrm{K}} \right)^{1/2} \mathrm{km\ s}^{-1}, \tag{3.9}$$

where μ_m is the mean molecular weight. For simplicity, this is taken as equivalent to the thermal velocity dispersion. In general, we can approximate: $v_{th} \sim 10^4 T^{1/2}$ cm s^{-1} where T is in kelvins. Therefore, the molecular viscosity becomes

$$v_m \sim 3 \times 10^{11} \left(\frac{n}{10^4\ \mathrm{cm}^{-3}} \right)^{-1} \left(\frac{T}{10\ \mathrm{K}} \right)^{1/2} \mathrm{cm}^2\ \mathrm{s}^{-1}. \tag{3.10}$$

The jet Reynolds number is defined here as

$$\mathcal{R}e_m = R_j U_j / v_m, \tag{3.11}$$

where R_j and U_j are the jet radius and speed, R_j representing the largest scale on which turbulent eddies could develop. The Reynolds number, being dimensionless, is relevant to flows on any scale. Therefore, to determine if jets are viscous, we only have to evaluate $\mathcal{R}e_m$. In the laboratory, viscosity is found to maintain laminar flows up to $\mathcal{R}e_m$ of order several thousand.

For neutral stellar jets, $R_j U_j \sim 10^{22} - 10^{25}$ cm^2 s^{-1} (see Table 1.2) and so the typical jet Reynolds number exceeds 10^{11}. Therefore, the flow will be unstable and fully turbulent boundary layers will be produced. To determine the thickness of the boundary layer or, indeed, if the turbulence should extend through the entire volume of the jet, several factors must be taken into account, such as the rate of opening of the jet, the motion of sound or magnetic waves across the jet, and the growth of fluid dynamic instabilities.

The H_2O and CO gas jets which drive cometary dust jets have been modelled as compressible fluids. Although the parameters are uncertain, we anticipate here the modelling results of Section 9.1.3. For a vent of diameter 10 cm, density $\rho = 10^{-2}$ g cm^{-3} and jet speed of 0.4 km s^{-1}, the mass outflow would be ~ 13 kg s^{-1}, which is consistent with present observations. The molecular viscosity can be written as

$$\nu_g \sim 1.5 \times 10^{-5} \left(\frac{\rho}{10^{-2}\,\text{g cm}^{-3}} \right)^{-1} \left(\frac{T}{250\,\text{K}} \right)^{1/2} \text{cm}^2\,\text{s}^{-1}. \tag{3.12}$$

Therefore, with a Reynolds number of $\sim 3 \times 10^{10}$, this is again many orders of magnitude above the values which would correspond to viscous flow for any size of vent.

At another extreme, the jets containing hot ionised plasma are magnetically active. The appropriate viscosity is given approximately by $r_L(r_L/\lambda_{ie})v_{th}$ where the Larmor radius, r_L, is defined in Section 2.1 and the mean free path λ_{ie} relates to the ion–electron collision frequency, $f_{ei} = 4 \times 10^{-6} n_i \Lambda / T^{3/2}$, where Λ is the Coulomb logarithm, taken as 30. It is straightforward to show that (De Young, 1991)

$$\nu_i \sim 5 \times 10^3 \left(\frac{n}{10^{-4}\,\text{cm}^{-3}} \right) \left(\frac{T}{10^8\,\text{K}} \right)^{-1/2} \left(\frac{B}{10^{-6}\,\text{G}} \right)^{-2} \text{cm}^2\,\text{s}^{-1}. \tag{3.13}$$

For large-scale extragalactic jets, $R_j U_j \sim 10^{30} - 10^{32}$ cm^2 s^{-1} and so the typical jet Reynolds number far exceeds 10^{27}. For the parameters derived from the microquasar SS 433 jet by Watson *et al.* (1986) and taking a kilogauss field, $\nu_i \sim 10^4$ and $\mathcal{R}e \sim 10^{16}$. It is clear that all detected jets are inviscid and that extremely narrow but highly turbulent boundary layers will be present.

In the case of a strong magnetic field, the magnetic Reynolds number describes the magnetic diffusivity, ν_{mag}, in a similar manner. The magnetic diffusivity measures the ability of the field to diffuse through the gas in the jet. It is thus simply inversely proportional to the conductivity, which for a hot ionised gas leads to $\nu_{mag} \sim 3.9 \times 10^{-14} T^{-3/2}$ cm^2 s^{-1}. Hence, the magnetic Reynolds number will again take extremely large values.

3.3　Magnetohydrodynamics

Magnetic and gravitational forces are an essential concern in the theory of jets. The realm of ideal magnetohydrodynamics (MHD) is a starting point that already carries with it high complexity. So we begin by assuming a single-component conducting fluid, ignoring for now ambipolar and magnetic diffusion including ohmic dissipation. In this case, the force exerted on a unit volume of fluid is

$$\rho \frac{d\mathbf{v}}{dt} = -\nabla p_t - \rho \nabla \Phi + \frac{\mathbf{j} \times \mathbf{B}}{c}, \tag{3.14}$$

where the current density is given by

$$\mathbf{j} = \frac{c}{4\pi} (\nabla \times \mathbf{B}). \tag{3.15}$$

The magnetic field thus results in the Lorentz force, derived from Maxwell's equations on assuming an electrically conducting medium and non-relativistic speeds. Applying a well-known vector identity, it is seen to be the sum of a magnetic tension, which acts to straighten field lines, and a negative gradient of the magnetic pressure, $P_b = B^2/(8\pi)$, which acts to push them apart:

$$\frac{1}{4\pi}(\nabla \times \mathbf{B}) \times \mathbf{B} = \frac{1}{4\pi}(\mathbf{B} \cdot \nabla)\mathbf{B} - \frac{1}{8\pi}\nabla(\mathbf{B}^2). \tag{3.16}$$

Note that electrostatic forces due to electric charges are not represented due to the condition of charge neutrality, and electrodynamic forces have been eliminated through Eq. (3.15). Hence, we do not need to refer explicitly to the electrical properties when the MHD approximation is valid.

Magnetic flux is a useful quantity which may be conserved along a jet path. In general, the field behaves as if frozen into the gas. To understand how a field develops, we add the induction equation, which expresses the change in magnetic field at a fixed location as

$$\frac{\partial \mathbf{B}}{\partial t} = \nabla \times (\mathbf{v} \times \mathbf{B}). \tag{3.17}$$

As implied by the name, a magnetic field can be induced through non-uniform fluid motions. Furthermore, the non-divergence initial condition

$$\nabla \cdot \mathbf{B} = 0 \tag{3.18}$$

must be satisfied everywhere. The magnetic problem is not over-specified by this fourth equation since it is intrinsic to the induction equation once satisfied at any instant. It simply ensures that magnetic monopoles are non-existent.

It is useful to define the magnetic flux through Eq. (3.17) by integrating over an arbitrary surface, S, which has $d\mathbf{S} = \mathbf{n}(dS)$ where $\mathbf{n}$ is the unit vector normal to the surface element dS. The flux is then defined as $\Phi_B = \int_S \mathbf{B} \cdot d\mathbf{S}$. The consequence is that the magnetic field can be represented by tubes of flux frozen into the fluid. As a tube changes cross-sectional area, A, the magnetic field strength does also, inversely proportionally to the flux tube area. That is, flux is conserved. In addition, material is free to move along the tubes without alteration of the field or flux. Therefore, in a simple non-rotating, non-sheared jet with a circular cross-section, we expect the axial component of the magnetic field to decrease as $B_z \propto 1/R_j^2$ and the azimuthal field to decrease as $B_\phi \propto 1/(R_j v_j)$, where R_j and v_j are the jet radius and axial speed.

The expansion of a jet, as well as many other properties, depends on two competing factors: the speed along the axis and the speed at which waves propagate across the jet. Hence, this ratio of speeds is central to all jet models. To determine how fast signals may propagate, a linear perturbation analysis of the MHD flow equations is undertaken (e.g. Shu, 1992). This yields the basic formula for sound waves of speed $c_s = \sqrt{(dp/d\rho)}$ in the hydrodynamic case (Section 3.1).

The adiabatic jet model presented below in more detail is one of the simplest we encounter. An adiabatic flow is defined as one in which we ignore all sources and sinks of energy, momentum and mass. Energy is converted between forms within a fluid parcel, the transfer occurring solely through expansion and contraction. Entropy is conserved and so the flow is also termed isentropic. The adiabatic law $p \propto \rho^\gamma$ then yields the adiabatic sound speed

$\sqrt{(\gamma p/\rho)}$. The pressure is also related to the internal energy per unit volume, e, by $p = (\gamma - 1)e$. Note that the constant of proportionality is not the same over the entire fluid, but that in a steady jet flow we may define 'streamlines' along which $p\rho^{-\gamma}$ is conserved where a streamline is defined as the path followed by an individual fluid element (Smith, 1994b).

Undertaking a linear analysis of the ideal MHD equations, three solutions are uncovered representing three distinct wave types. Firstly, non-dispersive waves move at the Alfvén speed, v_A, along the field lines or at a speed $v_A \cos \phi$ when propagating at an angle ϕ to the magnetic field direction. The fluid is set in motion transverse to both the wave direction and field direction but there is no associated compression. Here,

$$v_A = \frac{B}{\sqrt{4\pi\rho}} \tag{3.19}$$

which reflects that the transverse waves are supported by the tension of the magnetic field, rather like tensioned violin strings. Most notable, isolated Alfvén waves of any amplitude do not dissipate, and so may efficiently transport energy. In the case of protostellar jets or jets associated with young stars, likely parameters yield

$$v_A = 18.5 \left(\frac{B}{1 \times 10^{-3}\,\text{G}} \right) \left(\frac{n}{10^4\,\text{cm}^{-3}} \right)^{-1/2} \text{km s}^{-1}. \tag{3.20}$$

For extragalactic jets, indicative values yield

$$v_A = 0.05\, c \left(\frac{B}{1 \times 10^{-5}\,\text{G}} \right) \left(\frac{n}{10^{-5}\,\text{cm}^{-3}} \right)^{-1/2}. \tag{3.21}$$

The other two types of wave motion are magneto-acoustic. The magnetic pressure supports fast magnetosonic waves, which work in unison with the sound waves and so move at least as fast. In contrast, slow magnetosonic waves propagate more slowly than Alfvén waves due to counteracting thermal and magnetic pressures.

A hybrid wave relevant to jets is the torsional Alfvén wave. Such waves are central to models in which the magnetic field is continually twisted near the source, increasing the magnetic pressure, which drives an outflow along an axis (Shibata & Uchida, 1985). Torsional waves are also thought to be relevant to solar jets, since spicules exhibit rotation (Kudoh & Shibata, 1999).

For many of the jet classes the source of the energy is rotational, gravitational or solar energy. This energy is transferred into particle, thermal or magnetic energy before being converted into the bulk flow of the jets. Assuming a uniform jet flow, the ratio of bulk kinetic flow energy, E_K, to thermal energy, E_T, is proportional to the Mach number squared, $M^2 = v_j^2/c_s^2 = 2/(\gamma(\gamma - 1))E_K/E_T$ in the adiabatic case, while the ratio of bulk to magnetic energy, $E_B = B^2/8\pi$ is the Alfvén (Mach) number squared $M_A^2 = v_j^2/v_A^2 = E_K/E_B$.

3.4   Steady jets as potential flows

Full three-dimensional steady jets which are irrotational can be analysed by introducing a potential function for the velocity. Such a curl-free description can be used to elucidate mechanisms based on the de Laval nozzle and even centrifugally driven jets on the large scale where the angular momentum no longer influences the dynamics. We write the jet velocity $\mathbf{v} = \nabla\Psi$ in terms of the velocity potential. For an adiabatic flow with $p \propto \rho^\gamma$,

Euler's equation for steady flow, $\partial \mathbf{v}/\partial t = 0$, then yields energy conservation per unit mass along any streamline:

$$(\nabla \Psi)^2 + \frac{2\gamma}{\gamma - 1} \frac{p}{\rho} + 2(\nabla \Phi_G)^2 = U^2, \tag{3.22}$$

where U is Bernoulli's constant and Φ_G is the gravitational potential (Smith, 1994b). For an isothermal flow, it is straightforward to show that

$$(\nabla \Psi)^2 + 2c_i^2 \ln \rho + 2(\nabla \Phi_G)^2 = U^2, \tag{3.23}$$

where $c_i = p/\rho$ is the isothermal sound speed.

Mass conservation is written as

$$\nabla \cdot (\rho \nabla \Psi) = 0. \tag{3.24}$$

We do not introduce a stream function in this formulation. Instead, we eliminate p and ρ directly to yield the general potential equation for non-relativistic), steady compressible fluid flow:

$$(\gamma - 1)\left[U^2 - 2(\nabla \Phi_G)^2 - (\nabla \Psi)^2\right]\nabla^2 \Psi = (\nabla \Psi \cdot \nabla)(\nabla \Psi)^2. \tag{3.25}$$

in the adiabatic case.

This equation can in principle be solved on specification of the boundary conditions: the value across some section at the base of the jet and the nature of the interface. For a jet flow, we employ an interface with an ambient medium as a tangential discontinuity:

$$(\mathbf{n}) \cdot \nabla \Psi = 0. \tag{3.26}$$

In addition, we will assume a pressure balance across the interface with a specified pressure distribution for the environment (see Section 10.2).

3.5   Streamlines: rotating MHD flow

The interaction of accretion discs with the magnetospheres of young stars can produce X-celerators, X-winds and funnel flows. With the critical assumptions of two-dimensional axially symmetric steady-state flow, the problem can be formulated in terms of quantities that are conserved along streamlines, such as the Bernoulli integral, plus a partial differential equation called the Grad–Shafranov equation that governs the distribution of streamlines in the meridional plane.

The fundamental set of steady MHD equations adopted here appears in the literature in a non-dimensional form, with length and time scales corresponding to the radius, R_x and inverse angular frequency, $1/\Omega_x$ at some specified disc location from which magnetic field lines are rooted. Densities are expressed in terms of mass outflow, $\dot{M}_w/(4\pi R_x^3 \Omega_x)$ (Sakurai, 1985; Shu & Shang, 1997; Cai *et al.*, 2008).

The analysis is expediated by working in the co-rotating frame (see Section 9.3.2 for expressions in the fixed frame). In this field-rooted frame, we define non-dimensional cylindrical coordinates (ϖ, ψ, z) and velocity $\mathbf{u} = \mathbf{v}/(R_x \Omega_x)$ and re-write mass conservation as

$$\nabla \cdot (\rho \mathbf{u}) = 0. \tag{3.27}$$

With axial symmetry, we introduce a stream function $\phi(\varpi, z)$ which assures that mass is conserved:

$$\rho u_\varpi = \frac{1}{\varpi} \frac{\partial \phi}{\partial z}, \tag{3.28}$$

$$\rho u_z = \frac{1}{\varpi} \frac{\partial \phi}{\partial \varpi}. \tag{3.29}$$

The induction equation reduces to the flux freezing condition $\mathbf{u} \times \mathbf{B} = 0$ in the co-rotating disc plane from which $\mathbf{u}$ and $\mathbf{B}$ are assumed to be aligned. This implies that the magnetic field is proportional to the mass flux, $\mathbf{B} = \beta \rho \mathbf{u}$. Since $\nabla \cdot \mathbf{B} = 0$, we also require $\mathbf{u} \cdot (\nabla \beta) = 0$ so that the scalar $\beta(\phi)$ is conserved on a streamline. Hence, $1/\beta$ can be defined as the fixed *mass load* of a field line. In this model, the mass flux from a ring of the accretion disc depends only on R_x and is proportional to the magnetic flux. Its value will depend on the physical processes within and close to the disc surface.

To make progress, the three components of the momentum associated with streamlines are now treated. The steady-flow Euler MHD equation (3.14) in a rotating frame includes extra coriolis and centripetal terms:

$$\nabla \left(\frac{1}{2} \mathbf{u} \cdot \mathbf{u} \right) - \mathbf{u} \times (2\mathbf{e}_z + \nabla \times \mathbf{u}) = -\frac{\epsilon^2}{\rho} \nabla \rho + \frac{1}{\rho} (\nabla \times \mathbf{B}) \times \mathbf{B} - \nabla \mathcal{V}_{eff} \tag{3.30}$$

for an isothermal flow with $\epsilon = c_i/(R_x \Omega_x)$, where the field has also been non-dimensionalised so that $M_A^2 = 1/(\rho \beta^2)$, and where the gravitational potential modified by the effective centrifugal potential is

$$\mathcal{V}_{eff} = \frac{3}{2} - \frac{\varpi^2}{2} - \frac{1}{(\varpi^2 + z^2)^{1/2}} \tag{3.31}$$

(Cai *et al.*, 2008).

Firstly, we consider the toroidal component. The angular momentum of the gas is

$$\varpi (1 - \beta^2 \rho) u_\psi + \beta^2 \rho \varpi^2 = J(\Phi), \tag{3.32}$$

allowing for that part carried away by the Maxwell torque of the field. This expresses, as expected, that the total specific angular momentum along any streamline is conserved.

The component along the fluid velocity is found by taking the inner product of Eq. (3.30) with $\mathbf{u}$. We thus obtain the Bernoulli equation along streamlines:

$$\frac{1}{2} u^2 + \mathcal{V}_{eff} + \epsilon^2 \ln \rho = H(\Phi). \tag{3.33}$$

The third component of the Euler equation describes the momentum balance in the direction perpendicular to the poloidal field lines. This yields an equation which lies at the heart of MHD jet theory but which cannot be analytically integrated (see Pelletier & Pudritz (1992) for more on the derivation), the famous Grad–Shafranov equation:

$$\frac{1}{\varpi} \frac{\partial}{\partial \varpi} \left(\varpi \mathcal{A} \frac{\partial \Phi}{\partial \varpi} \right) + \frac{\partial}{\partial z} \left(\mathcal{A} \frac{\partial \Phi}{\partial z} \right) = \mathcal{Q}, \tag{3.34}$$

where $\mathcal{A}$, related to the Alfvén discriminant, is

$$\mathcal{A} = \frac{\beta^2 \rho - 1}{\varpi^2 \rho} = \frac{M_A^{-1} - 1}{\varpi^2 \rho}, \tag{3.35}$$

while Q incorporates terms which describe the effective gradient in gas pressure and toroidal flow and field effects which, in the end, determine how well the jet is collimated:

$$Q = \rho \left[\rho u^2 \beta \beta' + \frac{u_\psi}{\varpi} J'(\Phi) - H'(\Phi) \right],$$ (3.36)

with primes denoting differentiation with respect to Φ (Shu *et al.*, 1994). The left-hand side of the Grad–Shafranov equation involves an inertial term associated with poloidal motions and a magnetic tension term associated with the poloidal field. The right-hand side includes all other centrifugal terms, magnetic pressure gradients and hoop stresses.

The Bernoulli and Grad–Shafranov set of equations yields a partial differential equation of mixed type: elliptic before critical surfaces are crossed (where the flow speed equals characteristic wave speeds) and hyperbolic beyond. The computational difficulties are exacerbated by the locations of the critical surfaces not being known in advance. Nevertheless, a number of interesting properties relevant to disc wind models can be extracted (see Section 9.3.2).

3.6 Special relativistic flow

A description of relativistic hydrodynamics will embed within it the equations of classical hydrodynamics. It is here provided by considering the usual trinity of mass, momentum and energy. We first derive the general equations and then illustrate the properties by considering the potential flow of an ultra-relativistic gas.

Clear definitions are essential. We employ v as the fluid velocity with three components so that the Lorentz factor takes the usual form in a background frame, usually that of the jet source,

$$\Gamma = \left(1 - \frac{v^2}{c^2} \right)^{-1/2}.$$ (3.37)

The gas itself is relativistic (an ultra-relativistic gas is not assumed as yet although it is often implicit in the literature). In its rest frame, we may employ a perfect gas law in all regimes:

$$p = NkT = \rho c^2 / \eta,$$ (3.38)

where $N = m\rho/m$ is the particle number of rest mass m (baryon or lepton number, assuming no pair creation or annihilation) and η is defined as the dimensionless reciprocal temperature

$$1/\eta = \frac{kT}{mc^2}.$$ (3.39)

To be fully prescribed, a relationship is needed to determine the internal energy. This is well defined in the classical and ultra-relativistic limits, according to the physics of the problem. In between, however, in a mildly relativistic gas the Maxwell–Boltzmann distribution is modified in a manner which makes the relation between internal energy and pressure unknown. One approach introduces the function $G(\eta)$ as the dimensionless specific enthalpy. Then, the heat function per unit volume is

$$w \equiv e + p \equiv \rho c^2 G(\eta),$$ (3.40)

where e is the total internal energy density, $\rho(c^2 + \epsilon)$, p is the pressure, $N = \rho/m$ still, and ϵ is the specific internal energy.

One assumed form for the dimensionless specific enthalpy is $G(\eta) = K_3(\eta)/K_2(\eta)$ involving modified Bessel functions, K_i (Lanza et al., 1985) (see also Peacock (1981)). This specifically describes a single-component relativistic fluid. In the two limiting cases, it can be taken as

$$G(\eta) \rightarrow 4/\eta \qquad \text{for } \eta \ll 1 \text{ (ultra-relativistic limit)}, \tag{3.41}$$

$$G(\eta) \rightarrow 1 + 5/(2\eta) \qquad \text{for } \eta \gg \text{ (classical limit)} \tag{3.42}$$

for a monatomic gas. The sound speed is given by

$$\left(\frac{c_s}{c}\right)^2 = \frac{dG/d\eta}{[\eta G(dG/d\eta + 1/\eta^2)]}, \tag{3.43}$$

which approaches $c/\sqrt{3}$ in the ultra-relativistic limit. Finally, we can define a parameter analogous to the specific heat ratio,

$$\gamma = 1 + p/(\rho\epsilon) = 1 + \frac{1}{\eta(G - 1/\eta - 1)}, \tag{3.44}$$

which yields the correct limiting values including the ultra-relativistic values $e = 3p$ and $\gamma = 4/3$. Alternatively, Krautter et al. (1983) defined the parameter $y = \eta G(\eta)$ and derived

$$\gamma = \frac{\eta^2 + 5y - y^2}{\eta^2 + 5y - y^2 - 1}. \tag{3.45}$$

In general, the sound speed c_s can be defined as

$$c_s^2 = \frac{\gamma(\gamma - 1)pc^2}{(\gamma - 1)\rho c^2 + \gamma p}, \tag{3.46}$$

the jet Mach number as $M = v_j/c_s$, and the relativistic proper Mach number as

$$\mathcal{M} = M\frac{\Gamma}{\Gamma_s} \tag{3.47}$$

with $\Gamma_s \equiv 1/\sqrt{(1 - c_s^2/c^2)}$ (Konigl, 1980; Mendoza & Longair, 2001).

The relativistic Euler and energy equations are contained in

$$\frac{\partial T_i^k}{\partial x^k} = u_i\frac{\partial}{\partial x^k}(wu^k) + (wu^k)\frac{\partial u_i}{\partial x^k} + \frac{\partial p}{\partial x^i} = 0 \tag{3.48}$$

(following Daly and Marscher (1988)), where the stress-energy tensor is

$$T_{\mu\nu} = wu_\mu u_\nu + pg_{\mu\nu}. \tag{3.49}$$

Here, u^i is the four-velocity of the fluid, which satisfies

$$u_\mu u_\mu = -1, \tag{3.50}$$

ensuring motion at the speed of light within spacetime. Finally, $g_{\mu\nu}$ is the Minkowski metric tensor with $g_{11} = g_{22} = g_{33} = 0$, $g_{00} = -1$ and all off-diagonal components equal to zero.

The final 4-vector form is generated by multiplying Eq. (3.48) by u^i:

$$-\frac{\partial}{\partial x^k}(wu^k) + u^k\frac{\partial p}{\partial x^k} = 0. \tag{3.51}$$

Mass conservation is formulated in terms of the baryon number density N:

$$\frac{\partial}{\partial x^i}(Nu^i) = 0. \tag{3.52}$$

The three space components of velocity, v_1, v_2 and v_3, with bulk flow speed v, bring the above equations into the form

$$w\Gamma^2\left[\frac{\partial v_i}{\partial t} + (\mathbf{v} \cdot \nabla)v_i\right] + v_i\frac{\partial p}{\partial t} + \frac{\partial p}{\partial x_i} = 0, \tag{3.53}$$

$$\frac{\partial e}{\partial t} + (\mathbf{v} \cdot \nabla)e + w\left[(\nabla \cdot \mathbf{v}) + \Gamma^2\mathbf{v} \cdot (\mathbf{v} \cdot \nabla)\mathbf{v} + \Gamma^2\mathbf{v} \cdot \partial \mathbf{v}\partial t\right] = 0 \tag{3.54}$$

and

$$\frac{\partial N}{\partial t} + (\mathbf{v} \cdot \nabla)N + N\left[(\nabla \cdot \mathbf{v}) + \Gamma^2\mathbf{v} \cdot (\mathbf{v} \cdot \nabla)\mathbf{v} + \Gamma^2\mathbf{v} \cdot \partial \mathbf{v}\partial t\right] = 0. \tag{3.55}$$

We now explicitly consider a steady irrotational jet flow with an adiabatic ultra-relativistic gas. According to Eq. (3.44), $p \propto N^{4/3}$ and, using Eq. (3.41), $e = 3p$. Eq. (3.53) then reduces to the analogue of Bernoulli's equation:

$$(\mathbf{v} \cdot \nabla)\Gamma p^{1/4} = 0. \tag{3.56}$$

The relativistic Bernoulli equation is thus given by

$$w\Gamma/\rho = U^2, \tag{3.57}$$

where U is a constant along a streamline. The continuity equation becomes

$$(\mathbf{v} \cdot \nabla)p + 2p(\nabla \cdot \mathbf{v}) = 0. \tag{3.58}$$

Together, we can eliminate p to get

$$\Gamma(\nabla \cdot \mathbf{v}) - 2(\mathbf{v} \cdot \nabla)\Gamma = 0. \tag{3.59}$$

Writing the potential function as

$$\nabla \cdot \Psi = \Gamma_o\mathbf{v}, \tag{3.60}$$

where Γ_o is the Lorentz factor at the base of a jet, we then obtain the general potential equation for *relativistic* steady compressible fluid flow:

$$\left[c^2\Gamma_o^2 - 2(\nabla\Phi_G)^2 - (\nabla\Psi)^2\right]\nabla^2\Psi = (\nabla\Psi \cdot \nabla)(\nabla\Psi)^2. \tag{3.61}$$

This possesses the same form as the non-relativistic case, Eq. (3.61), except for a fixed effective adiabatic index of 2. We have replaced the Bernoulli constant by $\Gamma_o c$ as utilised by Smith & Norman (1981).

It can be remarked that it is only the equation of state which differs between the structure of these equations and those for non-relativistic flow. The dynamics of ultra-relativistic gas behave as if $p \propto \rho^2$. Therefore many of the results of classical gas dynamics can be immediately generalised to their relativistic counterparts.

More generally, Chiu (1973) developed a method for steady relativistic flow which Wilson (1987) employed to simulate jets. A transformation of variables was undertaken without the assumption of an ultra-relativistic gas:

$$\tilde{v}_i = cGu_i, \tag{3.62}$$

$$\tilde{\rho} = mN/G, \tag{3.63}$$

$$\tilde{p} = \rho c^2/\eta, \tag{3.64}$$

with i taking values 1, 2 and 3. The three equations for steady relativistic flow are then

$$\nabla(\rho\mathbf{v}), \tag{3.65}$$

$$\rho(\mathbf{u}\cdot\nabla)\mathbf{u} + \nabla p = 0, \tag{3.66}$$

$$\rho(\mathbf{u}\cdot\nabla)\left(h + \frac{1}{2}\mathbf{u}\cdot\mathbf{u}\right) = 0, \tag{3.67}$$

where h is the *relativistic enthalpy* defined as

$$h = \frac{1}{2}c^2G^2. \tag{3.68}$$

In this manner, we can employ the same mathematics as in a non-relativistic gas but with a different physical significance attached to the variables. For example, (Daly and Marscher, 1988) we can determine the structure of a relativistic jet as it expands and subsequently contracts on entry into a medium of low pressure.

3.7   Shock waves

The bright knots of emission observed in a large fraction of jets are the result of heating and particle acceleration in shock fronts. In this section, we present essential formulae which describe established steady shock waves. The relevance to jets will be discussed in later chapters. It should be remarked that shock waves are inevitable in supersonic jets for one major reason. That is, stopping or even just deflecting a flow moving faster than the sound speed causes pressure waves to pile up, since they cannot move upstream. The pressure waves combine and steepen until some form of viscous force across a narrow front can provide the pressure increase, and so generate a shock. The flow pattern must involve a shock configuration of some kind. In many circumstances, a distinct second shock will also be driven into the ambient medium.

However, shock waves are not detected just at the point of impact. Abrupt dissipation of energy, probably associated with luminous outbursts, from near the launch region triggers pressure jumps which drive steepening velocity waves down the jet. Long-term variations in launch velocity or density will cause smooth gradients in pressure. The pressure gradients are bound to steepen under almost all physical conditions, steepening until a form of viscosity in a narrow layer (the shock transition) mediates across the propagating disturbance. Shock waves may also easily form through the steepening of sound waves moving transverse to the jet flow. These waves are driven by transverse pressure gradients, possibly triggered by under-expansion and over-expansion as the pressure of the fast-flowing jet material sluggishly responds to changing surrounding conditions. This may generate standing reconfinement shocks rather than a travelling pattern. In addition, shock waves enter into the equation when a fluid instability is able to grow into a non-linear state before being advected.

There are many types of shock wave that are relevant to jets and should be introduced here. Before we do this, we will consider one critical zone that is relevant to all shock waves: the transition zone. In this zone, a steep gradient in thermal pressure that has arisen has finally become counterbalanced by the dissipation of bulk flow energy. In a collisional shock, particles moving with supersonic speeds into the shock front from the high-speed upstream side collide with thermalised particles. As a result, they also become thermalised themselves, and flow downstream with a subsonic speed. This transition layer is clearly just a few mean free paths in width. The shock cannot widen since signals cannot be sent upstream. The narrow transition, called the shock front, is often treated as a discontinuity, since the mean free path is much narrower than relevant physical scales. One exception is the molecular C-shock discussed below.

The specific physics which actually brings about the transition does not influence the jump conditions. The physics only determines how the transition itself is effected through the shock width and the nature of the frictional force responsible for decelerating inflowing matter. In ionised gas, the shock front physics is collisionless. In this case, plasma waves due to collective motions of charged particles are responsible for the thermalisation process. Hence, plasma shocks are also very thin, despite the low density often encountered.

The time necessary for a shock wave to form will be related to the rate at which a pressure wave can steepen. Once established, it could weaken as less jet material remains to be swept up, reducing the pressure difference across the front. In jets, this may occur because the source outburst is cut, diminishing the mass flux, or because available material has already been collected up into the knots of shocked material. Transverse expansion will also reduce the mass flux into a shock propagating down a jet.

In between, a shock may reach a steady state in the reference frame moving with the shock front. Mass, momentum and energy flowing downstream equate exactly to those in the approaching upstream flow. Note that the total flow of energy is conserved, because we assume that the shock is sufficiently narrow so that the amount of energy extracted via radiation is negligible. We will also take a planar flow, often consistent with the breadth of the shock surface in comparison to the narrow transition.

3.7.1 *Relativistic shock waves*

Various types of shock wave have been studied, the range of physical assumptions reflecting our uncertainty in the phenomena we detect. Here, we present a couple of scenarios which may be of general application.

Consider the case of a special relativistic flow with no magnetic field and pressure isotropy for the gas (e.g. see Blandford & McKee, 1976). We will analyse within the particular reference frame in which the upstream flow is normal to the shock plane, which we take as the x-direction. We assume that the flow has reached a steady state, as observed in the reference frame moving with the shock. In this frame, the conservation of mass is described by

$$[\Gamma \beta \rho] = 0 \qquad (3.69)$$

following the notation of Section 3.6 with $\beta = v_x/c$. The square brackets, [......], are used to indicate that upstream and downstream values of the quantities inside the brackets are to be equated. Upstream of the shock front is the undisturbed and uniform pre-shock (subscript 1) and the shocked downstream flow is referred to as the post-shock state (subscript 2). Energy conservation yields

$$\left[\Gamma^2\beta(e+p)\right] = 0. \tag{3.70}$$

Momentum conservation along the x-axis yields

$$\left[\Gamma^2\beta^2(e+p)+p\right] = 0. \tag{3.71}$$

The above two equations can be solved to give the pre-shock and post-shock flow speeds explicitly or, more neatly, in the form

$$\beta_1\beta_2 = \frac{p_2 - p_1}{e_2 - e_1} \tag{3.72}$$

and

$$\frac{\beta_1}{\beta_2} = \frac{e_2 + p_1}{e_1 + p_2}. \tag{3.73}$$

To this we supply an equation of state which can take the form

$$p = (\gamma - 1)(e - \rho c^2), \tag{3.74}$$

where γ can vary through the shock front due to the processing of the material.

For relativistic jets, we can suppose that the gas remains relativistic throughout the propagating shock, with $\gamma = 4/3$. The jump conditions then yield

$$\beta_1 = \left[\frac{1 + e_1/(3e_2)}{1 + 3e_1/e_2}\right]^{1/2}, \qquad \beta_2 = \frac{1}{3}\left[\frac{1 + 3e_1/e_2}{1 + e_1/(3e_2)}\right]^{1/2}, \tag{3.75}$$

while the compression ratio is

$$\frac{N_2}{N_1} = \left(\frac{3e_2}{e_1}\right)^{1/2}\left[\frac{1 + e_1/(3e_2)}{1 + 3e_1/e_2}\right]^{1/2}. \tag{3.76}$$

Therefore, material enters the shock front at close to the speed of light in a strong shock ($e_2 \gg e_1$), and leaves with a speed $\sim c/3$. Hence, the compression ratio is 3 although the proper density increases by an unrestricted amount, $3e_2/e_1$.

The magnetic weakening of relativistic shock waves has been discussed by Double *et al.* (2004). A planar magnetohydrodynamic relativistic shock which has reached a steady state is an extension to the above hydrodynamic case. We assume the magnetic field to possess components $(B_x, 0, B_z)$ in the x and z directions which will act to deflect the flow, which is initially in the x-direction, in the z-direction. We also assume pressure isotropy and, again for simplicity, sufficiently high upstream Alfvén speed so that magnetic plasma waves which scatter particles are frozen into the plasma and, importantly, so that we can treat $v_z \ll v_x$. Mass conservation is, as before,

$$[\Gamma\beta_x\rho] = 0 \tag{3.77}$$

with $\beta_x = v_x/c$. Energy conservation yields

$$\left[\Gamma^2\beta_x(e+p+\frac{B_z^2}{4\pi})-\Gamma(2\Gamma-1)\beta_z\frac{B_xB_z}{4\pi}\right]=0. \tag{3.78}$$

Momentum conservation is well approximated along the x-direction by

$$\left[\Gamma^2\beta_x^2(e+p+\frac{B^2}{8\pi})+p+\Gamma^2\frac{B_z^2-B_x^2}{8\pi}-\Gamma(\Gamma-1)\frac{\beta_z}{\beta_x}\frac{B_xB_z}{2\pi}\right]=0, \tag{3.79}$$

and along the z-direction by

$$\left[\Gamma^2\beta_x\beta_z(e+p+\frac{B^2}{8\pi})+(\Gamma-1)^2\frac{\beta_z}{\beta_x}\frac{B_z^2-B_x^2}{8\pi}-\Gamma\frac{B_xB_z}{4\pi}\right]=0. \tag{3.80}$$

To this, we add the conditions to ensure $\nabla\cdot\mathbf{B}=0$ and $\nabla\times\mathbf{E}=0$,

$$[B_x]=0, \quad [\Gamma(\beta_zB_x-\beta_xB_z)]=0. \tag{3.81}$$

and an equation of state. Double *et al.* (2004) obtain interesting results for the jump conditions by adopting the simple expression $\gamma_2=4/3+1/(3\Gamma_r)$ for the downstream adiabatic index, where Γ_r is the Lorentz factor corresponding to the relative velocity between the converging plasma frames.

3.7.2 Non-relativistic shock waves

A knowledge of oblique shocks with oblique magnetic fields is unavoidable in interpreting astrophysical jets. Associated with many jets we observe curved shock waves called bow shocks, both internal to the jet and driven by an advancing jet into the ambient medium. If the curved surface has a radius of curvature that far exceeds the thickness of the shock front, as is very likely, then each surface element can be treated as a planar shock. We also see jets which abruptly alter their track on the sky, which are interpreted as deflected jets. Unless the deflection angle is extremely small, an oblique shock is expected to directly generate the deflection, the shock wave itself stemming from the external deflecting entity and cutting across the supersonic flow. For brevity, the starting point here will be the general jump conditions derivable from the ideal MHD equations in which a narrow viscous transition layer separates pre-shock and post-shock media, as in the above sub-section. We also ignore gravity and radiation.

The objective here is to sketch the nature of the shocks and illustrate how the jump conditions can be derived in three commonly encountered situations. These are: hydrodynamic shocks (Case I), 'perpendicular' shocks (Case II, in which rather perversely the field is parallel to the shock front) and oblique shocks (Case III). The convention presented by Shu (1992) is followed here and, to take advantage of conservation laws, we determine flow speeds relative to the reference frame in which the flow is steady, the plane of the shock front.

We consider a flow with a speed $v_\perp$ transverse to a planar shock surface and speed $v_\parallel$ parallel to that shock front. Similarly for the magnetic field, we take the very general components $B_\perp$ and $B_\parallel$. That means we deal with six variables, including the pressure and density. Mass conservation is of course independent of the magnetic field strength (Eq. 3.77):

$$[\rho v_\perp]=0. \tag{3.82}$$

In addition, we rewrite Eq. (3.81) as

$$[B_\perp] = 0, \quad [v_\| B_\perp - v_\perp B_\|] = 0. \tag{3.83}$$

Using the above results, momentum conservation parallel to the shock front is

$$\left[v_\| - \frac{B_\perp}{4\pi \rho v_\perp} B_\| \right] = 0, \tag{3.84}$$

which, on using the above relations, leads to

$$\left[\rho v_\perp v_\| - \frac{B_\perp B_\|}{4\pi} \right] = 0. \tag{3.85}$$

Conservation of perpendicular momentum is expressed as

$$\left[\rho v_\perp^2 + P + \frac{B_\|^2}{8\pi} \right] = 0. \tag{3.86}$$

Finally, energy conservation yields

$$\left[\rho v_\perp \left(\frac{\gamma}{\gamma - 1} \frac{P}{\rho} + \frac{1}{2}(v_\perp^2 + v_\|^2) \right) - \frac{1}{4\pi}(v_\| B_\perp - v_\perp B_\|) B_\| \right] = 0. \tag{3.87}$$

In Case I, eliminating all magnetic field terms, the equations are solved to yield the hydrodynamic shock jump condition. We can solve for the downstream parameters in terms of their upstream values, replacing the shock itself by a discontinuity across which we specify the so-called Rankine–Hugoniot conditions. The problem is one-dimensional, with $v_\|$ a constant and $v_\perp$ taking the values v_1 upstream and v_2 downstream.

The Rankine–Hugoniot conditions depend only on the specific heat ratio and the upstream or shock Mach number $M \equiv v_1/c_s = (\rho_1 v_1^2/\gamma_1 p_1)^{1/2}$. The immediate post-shock values are $\gamma_2 = \gamma_1 = \gamma$ and $\rho_2 = S\rho_1$, where the compression ratio

$$S \equiv \rho_2/\rho_1 = v_1/v_2 = \frac{(\gamma + 1) \cdot M^2}{(\gamma - 1) \cdot M^2 + 2\gamma}. \tag{3.88}$$

To complete the solution,

$$p_2/p_1 = 1 + \left(1 - \frac{1}{S} \right) \cdot M^2 = \frac{2\gamma M^2 - (\gamma - 1)}{\gamma + 1} \tag{3.89}$$

and

$$T_2/T_1 = \frac{p_2}{S \cdot p_1}. \tag{3.90}$$

Note that $S > 1$ and in the strong shock limit, $M_1 \to \infty$, $S \to (\gamma + 1)/(\gamma - 1)$ which takes the value of 4 for a monatomic gas and 6 for a diatomic gas (e.g. H_2).

In Case II, the magnetic field has the greatest cushioning effect on the shock. The compression ratio is, after a lot of manipulation, given by the positive root of

$$2(2 - S)S^2 + \left[2\beta_p + (\gamma - 1)\beta_p M^2 + 2 \right] \gamma S - \gamma(\gamma + 1)\beta_p M^2 = 0, \tag{3.91}$$

where $\beta_p = 8\pi p_1/B_1^2$ is called the plasma beta. This yields $\rho_2/\rho_1 = B_2/B_1 = v_1/v_2 = S$ and

$$\frac{p_2}{p_1} = 1 + \gamma M^2 \left(1 - \frac{1}{S}\right) + \frac{1}{\beta_p}\left(1 - \frac{1}{S^2}\right). \tag{3.92}$$

In the general Case III, the oblique shock analysis, both the magnetic field and velocity are oblique to the shock front. It can most conveniently be treated by applying a transformation into a coordinate system that moves along the shock front such that the upstream velocity is parallel to the upstream magnetic field. In this new frame, $\mathbf{v} \times \mathbf{B} = 0$ throughout and the flow direction corresponds to the the field direction on both sides of the shock front. A cubic equation can then be derived for the compression ratio, the three roots existing when the shock speed exceeds the fast, Alfvén and slow MHD waves. The intermediate Alfvén shock is not really a shock but a field rotation, while both slow MHD and fast MHD shocks are compressive.

In an oblique slow shock, the parallel component of magnetic field can actually decrease across a shock front. Perhaps more observationally relevant to jet flows, a magnetic field which is perpendicular to the shock front and parallel to the jet can be bent at the front, deflecting the flow with it. This may seem counterintuitive at first since the flow could carry on as if the shock were hydrodynamic. However, for the hydrodynamic solution to occur would require a scenario in which imposed boundary conditions enforce that the flow cannot change direction. Such *switch-on shocks*, in which a field transverse to the flow is abruptly switched on, are particularly relevant to radiative shocks, since the cooling compressive layer which follows the front can further rotate the magnetic field through a large angle.

3.7.3 *Radiative shock waves*

Shock fronts are treated as adiabatic when the time that gas spends in traversing the front is small in comparison to the cooling time. After the front, the compressed hot gas that is produced will have a much shorter cooling timescale and will have a much longer time subsequently to cool in the downstream layer.

The gas observed in molecular and atomic jets of young stars is subject to strong cooling. Since the cooling time, of order of years, is considerably less than the dynamical time, which may be 10 000 years, the emission is expected to arise from restricted zones. These are radiative zones that follow the heating at the shock waves propagating down the jet. The flow in the radiative layer can be added on to the adjacent shock front. The simplest case to analyse is again one-dimensional hydrodynamics, in which the cooling time of the high-pressure downstream gas is considered to be less than the flow time transverse to the jet (or parallel to the shock front) which would remove it from the jet.

The radiative zone of a hydrodynamic shock begins from the downstream end of the shock front. In one dimension the continuity equation reduces to

$$\frac{\partial \rho}{\partial t} + \frac{\partial}{\partial x}(\rho v) = 0, \tag{3.93}$$

where $\rho(x,t)$ is the density and $v(x,t)$ is the velocity. In the steady state, the time derivative can be removed and we can integrate:

$$\rho \cdot v = \rho_2 \cdot v_2. \tag{3.94}$$

The same procedure can be performed on the momentum equation

$$\rho \frac{\partial v}{\partial t} + \rho v \frac{\partial v}{\partial x} = -\frac{\partial p}{\partial x} \tag{3.95}$$

to yield

$$p + \rho \cdot v^2 = p_2 + \rho_2 \cdot v_2^2. \tag{3.96}$$

For the energy equation, on substituting for the internal energy in

$$\frac{\partial}{\partial t} \left(\frac{\rho v^2}{2} + \rho \epsilon \right) + \frac{\partial}{\partial x} \left[\left(\frac{\rho v^2}{2} + \rho \epsilon + p \right) v \right] = -S_{rad} \tag{3.97}$$

we arrive at the steady-state equation:

$$v \cdot \frac{\partial \left[p/(\gamma - 1) \right]}{\partial x} + \frac{\gamma}{\gamma - 1} \cdot p \cdot \frac{\partial v}{\partial x} = -\Lambda(n, T, f_1, f_2, f_3, \ldots), \tag{3.98}$$

where the cooling function $-\Lambda(T, n, f)$ represents the energy loss per unit volume and is dependent on a combination of the temperature, T, the hydrogen nuclei density, $n \equiv n(H_{nuclei})$, and a range of atomic and molecular abundances, f_i, as well as the ionisation fraction and electron density.

Significant modification in the chemistry may occur in the radiative zone. The heating at the shock front can lead to dissociation of molecular hydrogen and other molecules in the radiative layer. Subsequent cooling, including that associated with the dissociation process itself, can result in the molecules re-forming well downstream. The molecular fraction is then a function of time and this alters the specific heat ratio. To incorporate this into the analysis, we note that the density of hydrogen nuclei can be separated into atoms and molecules: $n(H_{nuclei}) = 2 \times n(H_2) + n(H_{atoms})$. We define the molecular fraction as $f = n(H_2)/n(H_{nuclei})$, which may thus range in value from 0 for fully dissociated to 0.5 for fully molecular.

To find the specific heat ratio, we also include a helium content as $n(He) = 0.1 n(H_{nuclei})$ as the second most abundant element after hydrogen. The helium and hydrogen atoms possess three degrees of freedom corresponding to translation in the three spatial dimensions. On the other hand, molecular hydrogen possesses five degrees of freedom if rotationally excited, three translational plus two rotational. This will certainly be the case in the near-infrared jets discussed in Chapter 6. (Two additional vibrational degrees would be fully excited only at temperatures exceeding 6000 K; molecular hydrogen cannot exist long at such temperatures and is rarely observed above 3000 K.) It then follows that the specific heat ratio can be written (Suttner *et al.*, 1997):

$$\gamma = \frac{5.5 - 3f}{3.3 - f}. \tag{3.99}$$

Hence, if $f = 0$ (an atomic medium), then $\gamma = \frac{5}{3}$; whereas, if $f = 0.5$ (fully molecular), $\gamma = \frac{10}{7}$, which corresponds to a strong-shock compression ratio of 17/3.

These variables also lead to linear differential equations:

$$\frac{\partial \gamma}{\partial x} = \frac{-4.4}{(3.3 - f)^2} \frac{\partial f}{\partial x} \tag{3.100}$$

and

$$\frac{\partial f}{\partial x} = \frac{1.4 \cdot m_p \cdot (R - D)}{\rho_1 \cdot v_1}, \tag{3.101}$$

which leaves a single first-order equation which can be solved numerically without difficulty.

Unfortunately, the comparison of steady one-dimensional shock waves with observed shocks in jets must now be considered rather crude. Multi-dimensional shock configurations are observed and various instabilities are likely to operate. Ultraviolet radiation from the radiative zone of a fast shock will provide feedback in the upstream 'precursor' via ionisation. Many of these processes have been described with reference to the interstellar medium (Draine & McKee, 1993).

3.8   Non-ideal MHD and non-MHD

Ambipolar diffusion and reconnection are two important departures from magnetohydrodynamics. The description of jets in terms of ideal MHD corresponds to magnetic flux freezing: the gas does not cross the field lines. It is evident that ionised particles and the magnetic field are rigidly tied together since electromagnetic forces result in rapid particle gyration about field lines. Thus, in jet phenomena, the ions and electrons are expected to be tied to the magnetic field (except deep within accretion discs) and subject to the Lorentz force directly. In ideal magnetohydrodynamics, the neutrals and ions form a single co-moving gas component which is accelerated together by the Lorentz force and remains moving together because of frequent collisions.

In jets which are deeply embedded in clouds of high extinctions and which are not exposed to ionising radiation from the source, the ionisation level will be low. The density of ions may even be such that the frequency of ion–neutral collisions is insufficient to allow a single fluid description. In this case, the neutrals slip relative to the ion-magnetic gas and the resultant drag provides a frictional heating which may have important consequences. The drift, called ambipolar diffusion, has thus been invoked as a heating mechanism in jets.

Assuming very few ions, a balance of forces is quickly reached for them in which the Lorentz force and drag force are equal and opposite in direction. The drift speed, v_d, is then given by

$$v_d \chi \rho_n \rho_i = \frac{(\nabla \times \mathbf{B}) \times \mathbf{B}}{4\pi}, \tag{3.102}$$

where ρ_n and ρ_i are the neutral and ion densities, and χ is the ion-neutral momentum transfer coefficient. This yields a heating rate per unit volume of

$$\Gamma_{AD} = \frac{[(\nabla \times \mathbf{B}) \times \mathbf{B}]^2}{16\pi^2 \chi \rho_n \rho_i}. \tag{3.103}$$

For optical jets, the ionisation level may not be negligible. In this case,

$$\Gamma_{AD} = \frac{[\rho_n (\nabla \times \mathbf{B}) \times \mathbf{B}]^2}{16\pi^2 \chi \rho_i (\rho + \rho_i)^2}, \tag{3.104}$$

as shown by Shang *et al.* (2002). In these jets, ambipolar diffusion heating is weak because the drag coefficient is relatively large and X-rays may contribute to a relatively high level of ionisation.

Ambipolar diffusion is critical to the magneto-centrifugal wind model for jets, which relies upon the twisting magnetic field to remove and later collimate disc material. However, the ionisation level in accretion discs is expected to be extremely low. This implies that the magnetic field may slip through the disc material and the wind is then suppressed. Alternatively, the diffusion may help keep the magnetic field configuration in a steady open form necessary to drive the wind by inhibiting the effects of shearing and radial advection within the disc (Wardle & Koenigl, 1993).

The third influence of ambipolar diffusion is through supersonic ambipolar diffusion in molecular shock waves (Draine *et al.*, 1983). Without diffusion, shock waves are described as above in terms of a jump in parameters at a shock front followed by a subsonic radiative layer (i.e. a J-type shock). With diffusion, the ion–neutral friction provides the viscosity of the shock front. Although neutral–neutral collisions would produce a J-type shock, the low ionisation level means that ionmagnetosonic waves move far upstream very rapidly against the supersonic neutral flow. The ions will continue to drift upstream until they are eventually impeded through collisions with neutrals. Hence, the shock front is transformed into an extremely broad region in which the neutrals are gradually decelerated (Smith & Mac Low, 1997). While the ion momentum is of course extremely low, it is the momentum they transfer across from the moving magnetic field which is responsible for the high drag. A crucial consequence is a change in the gas properties which is now unlikely to be adiabatic since the time for radiative cooling is also extended. Thus, the sound speed is suppressed and the flow actually remains supersonic throughout a broad continuous transition, called a C-type shock.

The evidence for C-shocks in the context of molecular jets has not been directly based on resolved shock fronts as yet. This is because the shock thickness, L_n, is still narrow in comparison to other length scales, being approximately given by the Alfvén speed and the ion density:

$$L_n \approx \frac{v_A}{\chi \rho_i}$$

$$\approx 10^{15} \left(\frac{v_A}{10 \text{ km s}^{-1}} \right) \left(\frac{\chi}{9.2 \times 10^{13} \text{ gm s}^{-1} \text{ cm}^{-3}} \right)^{-1} \left(\frac{n_i}{10^{-23} \text{ cm}^{-3}} \right)^{-1} \text{ cm}$$

$$(3.105)$$

for possible but rather badly known jet fiducial values. The real three advantages of C-shocks in interpreting observed infrared jets are derived from the channelling of the shock energy into emission from cool gas, as observed (Giannini *et al.*, 2004), the higher shock speed limit before dissociation of molecules occurs, and the enhanced chemistry that takes place within the shock transition (Flower *et al.*, 1996).

Reconnection almost certainly takes place within solar jets and is expected, along with Alfvén waves, to be a significant heating mechanism. In this case, photospheric fluid motions continuously disturb the magnetic field configuration. However, if the jet is dominated by magnetic pressure, it will tend towards the lowest magnetic energy equilibrium configuration that is available to it subject to the constraints of magnetic diffusivity (Heyvaerts & Priest, 1984). If the typical reconnection time is short, then a force-free state will be reached in which $\nabla \times \mathbf{B} = \mu \mathbf{B}$ with μ a constant.

In astrophysical jets, it is possible for external influences to alter the field configuration (Konigl & Choudhuri, 1985). It may then be necessary for the magnetic field to relax to a new state and release the excess magnetic energy (Vekstein *et al.*, 1994). However, although qualitative progress is being made for reconnection in solar jets (Moore *et al.*, 2010), quantitative results have proved elusive.

3.9   Summary

It is now clear that considerable mathematics is behind any rigorous description of jet dynamics. Although we have introduced the basic ingredients in terms of fundamental equations, it remains the work of later chapters to knead these into models with a reasonable consistency. We first introduce one by one the family of jets and consider the combinations of physical and dynamical processes responsible for generating the observed complexity.

4

Observations of extragalactic jets

The largest known individual objects in the universe are carved out by extragalactic jets. These are giant radio galaxies which can have linear extents of up to several megaparsecs. It should be immediately remarked that this type of galaxy is not itself composed of stars, dark matter and an interstellar medium, but is instead created by events from within a compact engine embedded in the core of the actual galaxy. Twin jets extend from the nucleus of the galaxy to feed energy into two extended reservoirs of relativistic and magnetic energy. In common, the observed radio emission is due to the synchrotron process and the extended environments are optically thin. In contrast, the compact central radio *core* corresponding to the origin of the jets is optically thick, detected with a self-absorbed flat spectrum.

The study of extragalactic jets and that of radio galaxies are closely connected. The size and shape of the radio sources reveal how efficiently jets can penetrate through their host galaxy. Some radio sources are identified with or directly related to the propagating jets, while many others appear to be the subsequent overspill associated with the jets' abrupt termination. The latter sources consist of two distinct *lobes*, which contain *hot spots* through which the energy is transferred into the lobes.

The distinctive morphologies of radio galaxies have provided the most robust means of jet classification and we begin below by reviewing the classes. Such a classification scheme is, for this book, not just a means of book-keeping. Instead, a modification to the behaviour of the jets is required to produce each type of observed structure. There are many constraints to be placed on the nature of the underlying jets by considering various qualitative properties, such as their global symmetry and hot-spot structure. To go further requires a quantitative investigation of the sizes, luminosities, polarisation and spectral energy distributions. The information, however, remains limited and it has been a struggle to assimilate the morphological snapshots into an evolutionary picture.

The desire to detect the four components – core, jet, hot spot and lobe – at other wavelengths besides the radio has been satisfied in recent years. Infrared, optical and X-ray knots are associated with some of the brightest radio features. The results have removed some uncertainties in our knowledge and leave much less room for speculation.

Radio sources associated with quasars were at first treated as an unconnected class, since the radio emission was associated exclusively with the compact quasar cores and protruding parsec-scale jets. We attempted to put these sources into a distinct group of compact flat-spectrum objects as opposed to the extended steep-spectrum sources. The radio spectrum is evidently related to the compactness through synchrotron self-absorption (see Section 2.2).

This dichotomy has proved untenable, with extended sources found to possess compact cores and compact objects found to possess extended lobes. Furthermore, radio quasars have been found to possess extended structures which are similar to those of radio galaxies. The picture which emerges is that quasars are generally distant objects in which the core is made prominent by relativistic beaming (see Section 11.4). Our telescopes thus tend to pick out objects with jets propagating close to our line of sight. We discuss the compact jet structures in the following chapter, while we include below discussion of all extended radio sources.

Sizes and luminosities of extragalactic objects outside our local group are determined after calculating their distance. This requires us to assume a model for the expansion of the universe which includes a Hubble constant, H_0, to relate redshift and distance. The cold dark matter model assumed in this book will take parameters that are consistent with the concordance values with a non-zero cosmological-constant: $H_0 = 72$ km s^{-1} Mpc^{-1}, $\Omega_{rad} = 0$, $\Omega_{mat} = 0.27$, $\Omega_\Gamma = 0.73$, so that the sum of the critical densities for radiation, matter and dark energy, respectively, yield a flat universe with a total critical density $\Omega_{total} = 1$, and the deceleration parameter is $q_0 = -0.6$.

4.1 The morphological classes of radio galaxies

4.1.1 Edge-brightened and edge-darkened

Two main types of radio galaxy were recognised by Fanaroff & Riley (1974) after observations of all sources in the revised 3C catalogue with the Cambridge One-Mile Telescope. The distinction was based on the location of low-brightness regions relative to high-brightness regions. Those with low-brightness regions further from the associated optical galaxy than the regions of peak intensity were designated as Class I, and those with little emission beyond the region of peak intensity were called Class II radio galaxies. Since their work, this morphological division between FR Is and FR IIs has remained popular and relevant. Examples are displayed in Figs 4.1 and 4.2. The criterion is defined in terms of the ratio of the separation between brightness maxima to the entire source extent, although the degree of edge-darkening or edge-brightening is more precisely parametrised by calculating one-dimensional moments along the major axis (Jones, 1990).

The FR II radio galaxies are edge-brightened structures. The associated FR II jets or 'strong flavour' jets are one-sided, narrow and well collimated, although a weak counterjet is often discernible as illustrated by 3C 219 (Fig. 4.1) and the prototype Cygnus A (Fig. 1.2). Hot spots are located at the positions where the jets (whether visible or not) impinge on the external medium.

The FR I group were originally defined by their edge-darkened structure as exemplified by plume-type jets in 3C 31 (Fig. 4.2) and 3C 449 (Perley *et al.*, 1979). They are of low power and their linear size tends to be somewhat smaller. The radio structure is dominated by the emission from the core and the jets. The lobes gradually decrease in brightness with distance and the radio spectrum steepens at the outer extremities, consistent with ageing of the relativistic electrons. The FR I jets, sometimes called 'weak flavour' jets, are prominent, two-sided and broad with large spreading rates (Bridle & Perley, 1984). Twin jets are quite symmetrical on large scales. However, close to the nucleus there is a high asymmetry, with sometimes only one jet detectable. The asymmetry and collimation

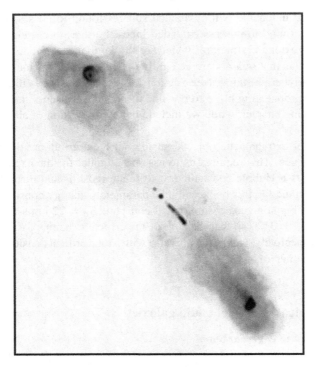

Fig. 4.1 A classical example of an FR II radio galaxy: 3C 219. This is a Very Large Array (VLA) image combining frequencies of 1.4 GHz and 1.6 GHz at 1.4" resolution. Note the bright, extended hot spots and the filamentary lobes. At a redshift of $z = 0.1745$, the luminosity distance is 822 Mpc and the total length is $\approx$ 550 kpc. The highly asymmetric jets, with only one knot seen here on the northern side, stem from the centre of a Seyfert galaxy, the dominant member of a compact cluster in Ursa Major. The radio flux measured at 178 MHz is 1×10^{27} W Hz^{-1} (Clarke *et al.*, 1992). Image courtesy of NRAO/AUI.

increase with the source power. On the other hand, the twin jets in some well-known FR Is display strong asymmetry (e.g. NGC 315 (Venturi *et al.*, 1993), and NGC 6251 (Fig. 1.5)). These properties are consistent with a decelerating twin-jet model in which a pair of inner jets are subject to relativistic beaming effects with typical jet speeds of $\sim 0.9c$). On the kiloparsec scale, however, the jets have been decelerated down to non-relativistic speeds (Laing *et al.*, 1999).

A high degree of polarisation is measured in the jets. This permits the direction of the component of the magnetic field in the plane of the sky to be constrained. It is found that the two-sided FR I jets are dominated by a parallel field close to the jet axis near the base (Bridle, 1984). It promptly transforms to perpendicular as a jet widens and becomes more symmetric. At high resolution, a parallel magnetic field is still detected at the periphery, suggestive of a field that has been sheared. However, recent modelling suggests that an ordered toroidal component combined with a disordered poloidal field could explain the data or, alternatively, even an entirely disordered field (Laing *et al.*, 2006). In contrast, in

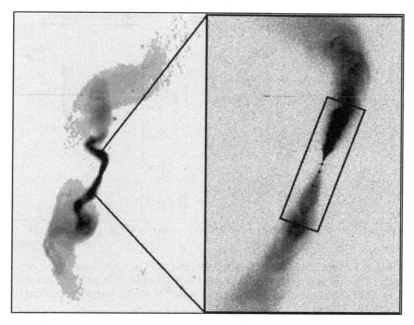

Fig. 4.2 An example of a type FR I radio galaxy: 3C 31. Left-hand panel: the VLA 1.4 GHz image of a 15" (300 kpc) twin-jet plume at 5.5" (1.9 kpc) resolution. Right-hand panel: VLA 8.4 GHz image of a 2" (40 kpc) field at 0.25" (85 pc) resolution. The rectangle within the right-hand panel demarcates the conical segment of the jets. This is a plumed radio galaxy at a redshift of $z = 0.0169$ (71 Mpc hosted by a dusty elliptical galaxy, NGC 383), the dominant galaxy within a prominent chain of galaxies. The radio flux measured at 178 MHz is 1×10^{25} W Hz^{-1}. Data presented by Laing & Bridle (2002b). Image courtesy of NRAO/AUI.

the one-sided jets associated with FR IIs, a parallel magnetic field component dominates over their entire lengths.

New species of radio galaxy have been identified after the accumulation of large numbers of sources from deep surveys. Within the second Bologna B2 sample are about 100 FR I low-luminosity radio sources ($10^{23} - 10^{25}$ W Hz^{-1} at 1.4 GHz). Some of these sources are quite intriguing. For example, there is an *intermediate* class of 'FR I/FR II' sources in which hot spots are located in the middle of the lobes. Furthermore, these sources possess FR I structure, but with one-sided jets which are well collimated. The two jets are asymmetric close to the core (within 10 kpc) but become symmetric further out. The structure is evident in the prototype B2 0844+31 displayed in Fig. 4.3. This suggests that these are transition objects in which an initially moderately relativistic jet ($0.5c$–$0.7c$) is decelerated to, at best, mildly relativistic speeds (Capetti *et al.*, 1995).

There is also a curious class of mixed FR I/FR II radio galaxies. These sources, called HYbrid MOrphology Radio sources or HYMORS (Gopal-Krishna & Wiita, 2000), possess an FR I morphology on one side of the host galaxy and an FR II morphology on the opposite side. One nice example is displayed in Fig. 4.4. The HYMORS studied possess radio powers

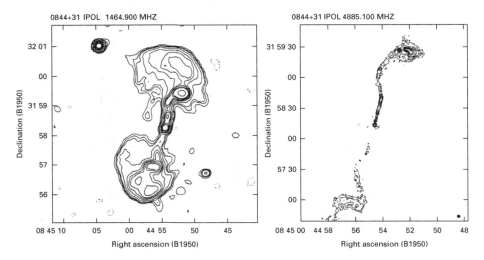

Fig. 4.3 A typical FR I/II radio galaxy: B2 0844+31, at 1.5 GHz and 10.9" resolution (left), and its jet structure at 5 GHz and 1.4" resolution (right panel). It lies at a redshift of $z = 0.0675$ and has a flux $P_{1.5\,\mathrm{Ghz}} = 1.3 \times 10^{25}$ W Hz^{-1}. Images taken from Capetti *et al.* (1995).

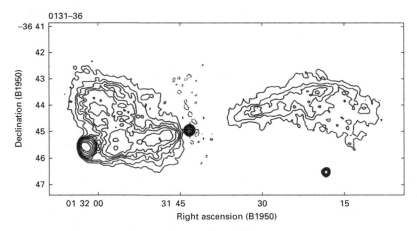

Fig. 4.4 An example of a hybrid radio galaxy, PKS 0131-36, observed at 4.9 GHz with the VLA. A knotty jet (not fully visible on this contour image) leads to a bright hot spot at the end of the eastern (left) lobe while a widening jet spreads into a diffuse western lobe. The host (S0 type) galaxy NGC 612 is at a redshift of $z = 0.029$, making the radio galaxy 390 kpc long with a flux of 2.2×10^{24} W Hz^{-1} at 5 GHz. Image from Morganti *et al.* (1993).

close to the FR I/FR II border and the associated nucleus can belong to a galaxy, quasar or BL Lac object.

In fact, the FR Is are a miscellany of sources. The vast majority of them are actually double-lobed but without hot spots, such as 3C 296 shown in Fig. 4.5. This radio

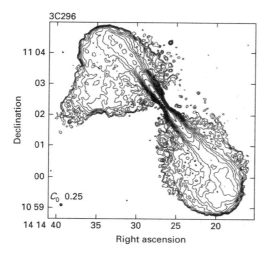

Fig. 4.5 A typical FR I lobed radio galaxy: 3C 296. VLA image at 1.5 GHz from Leahy & Perley (1991). The magnetic field (not shown) wraps around the lobe boundaries. At a redshift of $z = 0.029$, the size of the field shown is 178 kpc. Reproduced by permission of the AAS.

galaxy also displays a larger brightness asymmetry in the inner jets, consistent with the jets being initially of quite relativistic speed but slowing down further out. Plumed sources like 3C 31 turn out to be the exception. In the B2 sample, only 3% of low-luminosity sources are of the plumed type while 50% are double-lobed. The rest are tailed sources, discussed in Section 4.1.2, or 'naked jets', which are predominantly of low radio power ($P < 10^{23}$ W Hz^{-1}).

Note that the chance superposition of radio galaxies is rare. However, a fascinating case of interacting radio galaxies is associated with the dumbbell galaxy NGC 1128 at the centre of an Abell cluster of galaxies. The radio object is 3C 75, displaying unique interacting jets and plumes emanating from the double nucleus (Fig. 4.6).

Extragalactic radio sources cover a wide range in luminosity and the division of the classes is strongly related to it. The strongest sources all have edge-brightened morphology, while weak sources may be edge-brightened or edge-darkened. The dividing line was first drawn at $P^*_{178} = 1.2 \times 10^{33}$ erg s^{-1} Hz^{-1}, which is traditionally expressed in terms of watts: $P^*(178\,\mathrm{MHz}) = 1.2 \times 10^{26}$ W Hz^{-1} (upon converting to a Hubble constant of 72 km s^{-1} Mpc^{-1}), where $P(178\,\mathrm{MHz})$ is the flux at 178 MHz (see Section 2.8). Most radio galaxies are of lower power and are, therefore, FR Is.

To understand what these figures mean, the flux can be converted into the total power of the source, i.e. the radio luminosity. Assuming a synchrotron spectrum with a power-law index $\alpha > -1$, extending between sharp cut-offs at lower and upper frequencies, ν_1 and ν_2, the integrated power is

$$L = \int_{\nu_1}^{\nu_2} P^*_{178}(\nu/178\,MHz)^\alpha d\nu, \tag{4.1}$$

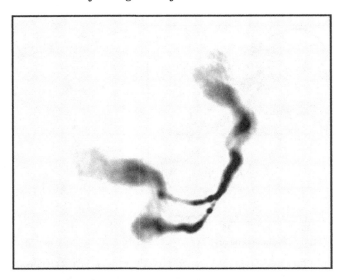

Fig. 4.6 The quadruple jets in 3C 75 (also called the Mating Dance by Waldrop (1985)). The radio jets are launched from the vicinity of two supermassive black holes located at the two bright spots in the image, separated by 8 kiloparsecs. These black holes are in the dumbbell galaxy NGC 1128 at a redshift of $z = 0.0231$. This is a VLA image with 0.8" resolution at 8.5 GHz. Credit: Owen *et al.* (1985), NRAO/AUI/NSF/NRL. Reproduced by permission of the AAS.

which yields the critical dividing luminosity of

$$L^* = 2.1 \times 10^{41} \frac{1}{1+\alpha} \left[\left(\frac{\nu_2}{178\,\text{MHz}} \right)^{1+\alpha} - \left(\frac{\nu_1}{178\,\text{MHz}} \right)^{1+\alpha} \right] \text{erg s}^{-1}. \qquad (4.2)$$

The spectral indices of most radio galaxies lie in the range -0.7 to -0.9. Taking $\nu_2 = 10\,\text{GHz}$ and $\alpha = -0.8$ yields a break luminosity $L^* = 2.1 \times 10^{42}$ erg s^{-1}.

The fraction of sources with jets decreases with increasing source power, with the less powerful radio galaxies known to have more symmetric jets. The radio cores in FR Is are also weaker than in FR IIs, although a well-defined break power is not evident. The division, however, also depends on the optical luminosity, as discussed by Owen & White (1991). The radio power at which the transition from FR I to FR II morphology occurs is given by $P_{178}^* \propto L_{opt}^{1.65}$, approximately (Gopal-Krishna & Wiita, 2001). That is, the more luminous the host galaxy is, the stronger the radio source must be in order to produce a classical double FR II. However, it should be noted that even quasars can possess FR I structure (Blundell & Rawlings, 2001; Heywood *et al.*, 2007).

The reason for the difference between the two jet types can be attributed to velocity or Mach number at the base of the jets. The FR II jets are relativistic and highly supersonic, while the FR I jets are transonic or mildly supersonic (Mach number < 5). The FR I jets are prone to fluid instabilities leading to turbulence and deceleration (Bicknell, 1984).

4.1.2   Wide-angled tails

Wide-angle tail radio galaxies (WATs) are an uncommon but fascinating class of radio source. They are characterised by twin, well-collimated radio jets which may extend

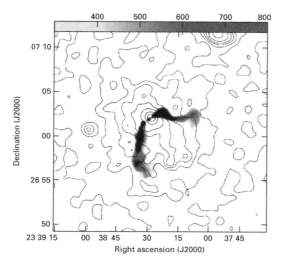

Fig. 4.7 A wide-angle tail radio galaxy: 3C 465. The greyscale is a 327 MHz VLA image. The contours show the X-ray emission associated with the Abell cluster of galaxies, A2634, observed with the ROSAT PSPC, convolved to 60" (35 kpc resolution). The size is 354 kpc at a redshift of $z = 0.0293$, corresponding to a luminosity distance of 125 Mpc. (Eilek & Owen, 2002). Reproduced by permission of the AAS.

for tens of kiloparsecs before abruptly flaring into long, diffuse plumes or tails. The original feature which distinguished the class is the bent plumes, often bent sharply although not through a large angle. Well-explored examples are 3C 130 (Hardcastle, 1998) and 3C 465, the latter displayed in Fig. 4.7. An established fact is that WATs are always associated with a cluster dominant galaxy, at or near the cluster centre. Their radio luminosity places them near the FR I/FR II luminosity break.

Despite their rarity, WATs have attracted considerable attention, since their distinctive properties provide plenty of nourishment for modellers. They are superb examples of how jets interact with their surroundings: the interstellar and intracluster media. Studies have concentrated on three aspects: the jet propagation, the jet disruption and the subsequent plume.

A jet with a surrounding sheath structure (Katz-Stone *et al.*, 1999) is invoked to explain a wide variety of statistical results including the brightness asymmetries, core prominence, polarisation structures and depolarisation. In past observations, the jets often appeared to have a gap between the core and the brightening on the scale of several kiloparsecs. Higher-sensitivity observations, however, demonstrate that a well-collimated jet pierces through the gap before flaring out into a plume-type jet with a surrounding sheath.

The speed of WAT jets can be predicted if the assumptions are made that a set of sources sample all possible angles to the line of sight and that the brightness of the jets is intrinsically symmetric. Attributing the entire asymmetry in brightness (the jet-sidedness ratio) to relativistic beaming yields jet speeds in the range $\beta c \sim 0.3c-0.7c$ (Jetha *et al.*, 2006). There is no direct independent evidence for relativistic speeds in these jets. Nevertheless, it seems plausible that beaming accounts for a substantial fraction of the jet–counterjet asymmetry.

In support, the two sources that show variable radio cores (an indication of beaming) also show comparatively one-sided jets.

What causes a WAT jet to disrupt? An important clue is the detection of a bright feature called a 'flare point' in many sources. It lies in the vicinity of where the jet terminates, i.e. near the base of some plumes. These are sometimes referred to as hot spots, although they are insufficiently resolved transverse to the jet to reveal their structure. This feature resembles the terminal hot spots found in FR IIs, suggesting that the jet also terminates in a strong shock and that WATs are close relations to FR IIs rather than FR Is.

For example, the WAT 3C 130 possesses a pair of relatively symmetrical, well-collimated inner jets, one of which exhibits a bright, compact, sub-kiloparsec feature at the northern flaring point (Hardcastle, 1998). This feature resembles the terminal hot spots found in FR IIs, suggesting that the jet also terminates in a strong shock. The absence of a similar feature in the southern plume, and some other WAT jets, may be related to the particle acceleration mechanism associated with the shock.

The southern jet of 3C 130, as well as some other WAT jets, does not display signs of abrupt jet termination. Instead, the jets penetrate into the plumes without disruption for some distance. The suggestion is that termination shocks are observed only in the cases where the jet happens to intersect with the interface of the plume and ambient medium. Otherwise, the jet may simply gradually merge into the plume flow.

The distance of the base of the plume from the galactic centre is related to the temperature of the gas in the host cluster (Hardcastle & Sakelliou, 2004). Although the sample remains quite small, shorter jets are found systematically in hotter clusters. This provides evidence that the structure of WATs is determined by their cluster environments. The model in which the plumes are generated when the jet crosses the interstellar/intracluster interface (Smith, 1982) then implies that the interface radius is smaller for hotter clusters. The model has the attractive feature that it predicts a scale length for the transition of tens of kiloparsecs, as observed (Hardcastle & Sakelliou, 2004).

The mechanism which disrupts the jets is still not clear, although several proposals have been put forward. In the 'cocoon-crushing' or 'failed-FR II' model, WATs form after the lobes of FR IIs are expelled by buoyancy forces, exposing the jet to the external medium (Hardcastle, 1999). Disruption would occur only for weak FR IIs, in which the expansion speed, driven by the internal lobe pressure, may become subsonic and the internal and external pressures become comparable while the source is still close to the cluster centre. With a high external pressure and sound speed in rich, hot clusters, buoyancy may remove the lobes earlier in the life of the parent FR II, resulting in shorter jets.

Recent findings demonstrate that large-scale dynamics in the host cluster are critical to the jet bending. Indeed, the host clusters of bent sources often show an X-ray structure that is elongated in the same direction as the WAT tails (Gomez *et al.*, 1997). That is, the thermal gas is asymmetric and aligned towards the direction of the bending. It is now believed that these clusters are undergoing cluster–cluster mergers, resulting in large-scale flows of hot gas. The jet and plume bending in merging clusters are, therefore, diagnostic of the 'stormy weather' in the intracluster medium and, consequently, of the evolving gravitational potential resulting from the merger (Burns *et al.*, 2002). In support, numerical simulations have illustrated the complex and changing cluster environments in which many radio galaxies live and evolve (Burns *et al.*, 2002). Groups and clusters of galaxies form at the intersections of filaments where they accrete gas and dark matter. The accretion process generates shocks and turbulence within the transonic bulk flow. The result is an unstable climate within which radio sources are easily distorted.

4.1.3 Narrow-angled tails

Narrow-angle tail radio galaxies (NATs) appear as twin bright jets issuing from the active nuclei (Ryle & Windram, 1968). Both jets curve through angles which can exceed 90° to generate a U-shaped structure. The two resulting lobes merge into a single extended tail which may extend several hundred kiloparsecs. Hence they are also called head-tail sources (Miley *et al.*, 1972). The classic example is the horseshoe shape associated with the galaxy NGC 1265, shown in Fig. 4.8. Note that projection effects cause some confusion between observed NATs and WATs and their physical counterparts. The jets are quite knotty and bends, wiggles and kinks provide a wealth of structure. The magnetic field runs parallel to the jet and is highly polarised, typically 10 – 30%.

NGC 1265 is the most prominent head-tail galaxy in the Perseus cluster (Fig. 4.8). The host galaxy lies on the outskirts of the area where X-ray emission is detected (Schwarz *et al.*, 1992) and moves with a large radial speed of 2300 km s^{-1} relative to the mean value for the cluster. Milliarcsecond very long baseline interferometry observations show two jet-like features on the parsec scale (at a distance of 78 Mpc, 1 milliarcsecond is 0.37 pc) and the jets are well aligned with the kiloparsec-scale structure (Xu *et al.*, 1999). The two kiloparsec-scale jets emerge from the galaxy and are bent around in the intracluster medium to merge into the perpendicular tail. On larger scales, the remarkable tail itself bends around almost through a full circle of diameter 280 kpc (Fig. 4.9).

Many WAT and NAT jets are surrounded by coaxial sheaths of radio emission (Katz-Stone *et al.*, 1999) For the head-tail source NGC 1265, the radio sheaths are tens of kiloparsecs long. The radio spectra are steeper than in the jets and also tend to be more highly polarised than the flat components. The sheath emission could arise from relativistic electrons which

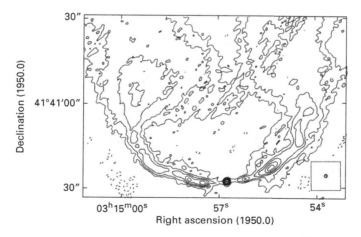

Fig. 4.8 The prototypical narrow-angle tail radio galaxy 3C 83.1B belonging to NGC 1265 at $z = 0.0183$. Note that this is just the inner arcminute jets which feed the much larger source shown in Fig. 4.9. The 4.87 GHz (6 cm) image was presented by O'Dea & Owen (1986).

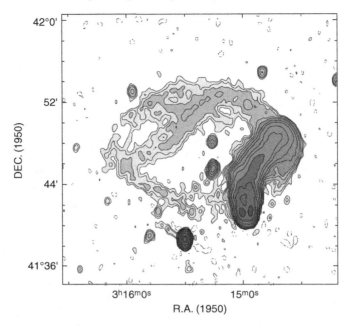

Fig. 4.9 A large-scale view of the narrow-angle tailed radio galaxy 3C 83.1B (NGC 1265). The panel shows the radio tail after the twin jets have merged and going on to bend through 360° (Sijbring & de Bruyn, 1998). Reproduced by permission of the AAS.

have diffused out of the jet into a static external medium. A thick boundary layer could develop within which particles are transferred via turbulent entrainment (Hanasz & Sol, 1996). Alternatively, a two-component jet flow can be hypothesised. For example, an electron–positron jet directly from a black hole could be surrounded by a slower electron–proton wind from the surrounding accretion disc.

The NATs are usually found in rich galactic clusters and the host galaxy has a large motion with respect to the embedding cluster. In fact, the distorted structure of all FR Is indirectly reveals the presence of a dense intracluster medium (Rudnick & Owen, 1976). Either the intracluster gas is in hydrostatic equilibrium within the cluster potential, and the galaxies are moving in the cluster potential, or the cluster gas is violently disturbed during cluster–cluster mergers. In either case, the jets appear to have been swept back by the deflecting force of the incident dense intracluster medium. The extra pressure arises from the relative motion of order 1000 km s^{-1} between the host galaxy and the intracluster medium.

On the other hand, some NATs are found in poor groups where both the average density of the intracluster medium and the velocities of the galaxies are supposedly too low to generate sufficient ram pressure (Venkatesan *et al.*, 1994). However, if these poor clusters are dynamically young and still collapsing, higher speeds are expected. Curved jets would be generated by the ram pressure corresponding to a relative speed of $v_{gal} = 600$ km s^{-1} through a medium of density $\rho_{icm} = 10^{-4}$ cm^{-3} (Venkatesan *et al.*, 1994)

The bending of the jets in head-tail sources has two possible origins. Direct ram pressure of the intracluster medium may divert the jets after they emerge from the centre of the galaxy

(Begelman *et al.*, 1979). The jets are supersonic and the trajectories can be calculated with the ram pressure $p_{ram} = C_r \rho_{icm} v_{gal}^2$, where C_r is a constant of order unity. Alternatively, the jets may be bent indirectly by the thermal pressure gradient within the galaxy's interstellar medium, which is itself set up by the motion of the intracluster medium past the galaxy (Jones & Owen, 1979). This latter model requires transonic jet speeds in order that the pressure gradient can be effective. The pressure gradient across the interstellar medium also implies the presence of a gigantic bow wave and a galactic wake, whose turbulence may help power the large-scale structure of the tailed radio source.

4.1.4 Classical doubles

The classical double or FR II radio galaxies are powerful but comparatively distant. Hence, with less resolution, detailed studies are not common. The lobes tend to be fairly symmetric in both size and luminosity. They are usually co-linear and a straight line can often be drawn through the hot spots and the core, although there are exceptions. They are generally associated with normal giant elliptical galaxies, which are considerably fainter than first-rank galaxies in rich clusters (Leahy & Williams, 1984). In fact, most classical doubles are not found in rich clusters although Cygnus A (Fig. 1.2) is one of the exceptions.

The radio lobes extend back towards the radio core, in some cases forming a long radio bridge (Leahy & Williams, 1984). The radio spectra of bridges tend to be progressively steeper closer to the source, consistent with synchrotron ageing of the relativistic electrons after being ejected from the jet at the hot spots. Adiabatic expansion and inverse Compton processes also contribute to the loss in lobe energy. The flatter spectrum of the jets and hot spots ($\alpha \sim -0.5$ to -0.9) in comparison to the lobes ($\alpha \sim -0.7$ to -1.2) is also consistent with the picture that the hot spots convert some of the bulk jet energy into relativistic electrons, in addition to compressing the magnetic field, before the high pressure spills the plasma into a cocoon which surrounds the advancing jet (e.g. Kharb *et al.*, 2008).

In early studies of FR IIs, jets were suspected but often not detected, requiring a high dynamic range and spatial resolution to be found. The jets and cores actually have high luminosity compared to FR Is, but are not prominent since the lobes are so powerful. Now, jets are detected in over 70% of nearby radio galaxies on the kiloparsec scale. The jets are almost exclusively one-sided, i.e. a flux 'sidedness' ratio above 4.

The physical properties of the jets which supply classical radio galaxies are only beginning to be established. The power is typically $10^{44} - 10^{46}$ erg s^{-1} which appears to be roughly constant over a lifetime of a few $\times 10^6 - 10^7$ years (O'Dea *et al.*, 2009). At relativistic speeds, this implies a total mass flow of $\sim 10^6 - 10^7 M_\odot$. However, the speed of the observed jet material has been hard to quantify. An indirect measure is made by attributing the sidedness to relativistic beaming. It can be safely assumed that the low radio frequency luminosities of the lobes should be intrinsically unaltered by the effects of beaming, so defining an unbiased sample. The prominence of the jets on kiloparsec scales can then be interpreted with a simple relativistic beaming model which implies a bulk jet speed of $\sim$0.6–0.7c, and a very efficient jet with a kinetic power which is approximately proportional to the source luminosity (Hardcastle *et al.*, 1999).

The internal structure is similar in the majority of radio jets whose paths can be traced along most of the source extent (e.g. 3C 47, 3C 109, 3C 200 (Fig. 4.10, 3C 275.1, 3C 334). The jets are not smooth but knotted, with a magnetic field aligned with the jet axis (Bridle

& Perley, 1984). The opening angles are less than 6° with a mean value of just 3°. Each jet appears to follow a straight path from the core to the end of the lobe where it bends sharply before terminating, often in a bright hot spot embedded on the side wall of the lobe. In no case does the jet follow a straight path from the core to the hot spot (Gilbert *et al.*, 2004). In two cases (3C 200 and 3C 334), at the point where the jet bends, a secondary jet-like structure is apparent on the outside edge of the bend.

4.1.5   Lobe-dominated quasars

Deep surveys for radio sources are dominated by high-luminosity extended structures around quasars (Stocke *et al.*, 1985). The structure of these radio-loud quasars closely resembles the powerful FR IIs (see Figs 4.10 and 4.11). The jets are again one-sided with a high collimation and an aligned magnetic field. One difference is that the jet luminosity is higher and, therefore, the jet is more easily detected. In addition, the core luminosity is also higher than that associated with radio galaxies of the same radio luminosity.

A radio quasar is identified as a source with an optical luminosity dominated by a point source rather than a host galaxy. They are, by a factor of 2, systematically smaller in extended radio structure than radio galaxies of similar power at a similar redshift. In addition, lobe structures show more distortion or bending than generally found in radio galaxies. A significant fraction (40%) of the FR II quasars are bent by more than 10° (de Vries *et al.*, 2006). This may be interpreted as a combination of three mechanisms – jet re-orientation (e.g. wiggling or precession), interaction with either the ambient gas (intergalactic or intracluster medium), and a large amplification through projection close to the line of sight.

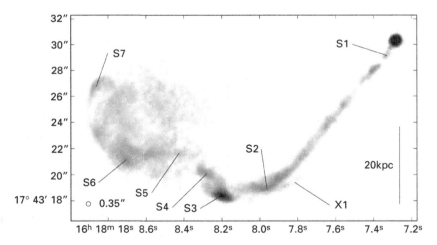

Fig. 4.10 The southern jet of 3C 334, associated with a double-lobed quasar at a redshift of $z = 0.555$. The one-sided jet is quite straight for approximately three-quarters of its length before the ricochet to the east (left). The jet apparently terminates at a hot spot at the location S3. From S3, a trail of emission (S4, S5) then curves back again towards a compact region (S6) at the edge of the lobe. The trail then arcs around to another bright feature at S7. In contrast, there is a distinct gap between the lobe emission and the core in the northern lobe. Image from Gilbert *et al.* (2004). Superluminal motion at 2.2c has been detected in the core.

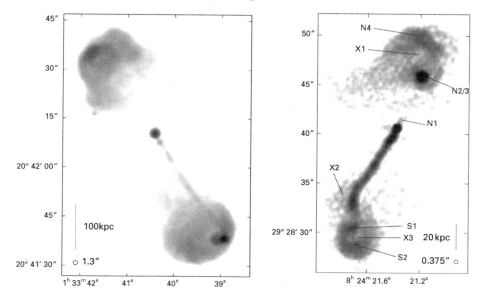

Fig. 4.11 Two examples of powerful radio sources. *On the left*, 3C 47 displays an S-shaped symmetry. It is associated with a quasar at a redshift of 0.425; 1 arcsec corresponds to 5.5 kpc. A superluminal motion of 4.3 c was found in the compact core (Gilbert *et al.*, 2004). *On the right*, 3C 200, at a redshift of 0.458, displays a prominent one-sided jet. However, the associated optical object is a narrow-line radio galaxy possibly implying that the axis is at a large angle to the line of sight. The distorted lobe morphology is also more typical of a quasar. The jet is polarised (~30%) with the magnetic field aligned with its length. After the bending, however, the polarisation drops to ~10% and the magnetic field direction turns to be transverse to the jet (Gilbert *et al.*, 2004).

A further difference is that radio quasars possess broad emission lines in their optical spectra whereas most powerful radio galaxies display only narrow emission lines. Of course, radio galaxies may also possess a compact nuclear source and associated broad line region, but they are then obscured from view. The evidence for orientation-dependent obscuration leads to the unification of the set of objects which appear as radio galaxies and quasars (Barthel, 1989). It predicts the lack of quasars viewed at large angles to the radio axis. These must appear as radio galaxies. However, not all classical radio galaxies possess hidden quasars since infrared observations with the *Spitzer* space telescope have uncovered a number which do not possess the obscuring dust (Ogle *et al.*, 2006).

4.1.6 *Relaxed doubles*

Some radio galaxies are very poorly collimated, providing difficulties for jet models which are thought to generate highly collimated flows. Examples of these 'relaxed' or 'fat' double sources (Owen & Laing, 1989) are 3C 310, 3C 401, Fornax A and Hercules A, the latter two shown in Fig. 4.12 and Fig. 4.13. All of these sources are largely devoid of hot spots and, often, of jets but possess many warm spots and filaments. Their luminosity places

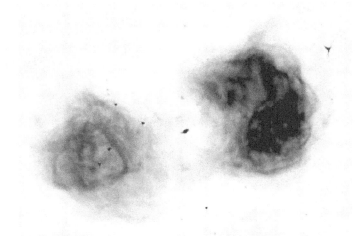

Fig. 4.12 A fat or relaxed radio galaxy, Fornax A, imaged with the VLA at 1.5 GHz with a resolution of 14", corresponding to 2.2 kpc. It is one of the nearest radio galaxies at a distance of 23 Mpc with radio lobes spanning 33". It is identified with the ninth-magnitude cD galaxy, NGC 1316, in a poor cluster (Fomalont *et al.*, 1989). Copyright NRAO. Reproduced by permission of the AAS.

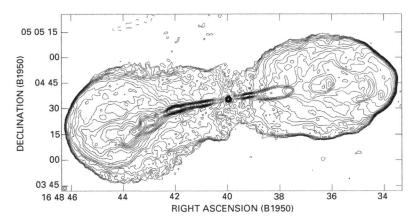

Fig. 4.13 The fat double Hercules A. A VLA 1440 MHz image with a beam size of 1.4", as shown in the lower left-hand corner, corresponding to 3.5 kpc. Contours are logarithmic. It is one of the brightest radio galaxies in the sky and is identified with the central cD galaxy of a cluster. The radio jets and the major axis of the cD galaxy are aligned to within $\sim 10°$. Note the bulbous lobes. The jet has a flatter spectrum suggesting that renewed activity may be occurring. At the low redshift of $z = 0.154$, the linear size and width are 490 kpc and 180 kpc. Its radio flux density at 178 MHz is 1.5×10^{27} W Hz^{-1} sr^{-1} and the total radio luminosity is $\sim 3 \times 10^{44}$ erg s^{-1}. Copyright NRAO (Gizani & Leahy, 2003).

them near the FR 1–FR II transition and they are found in clusters. Their radio spectra are quite steep, which all suggests that they are dying radio sources with no fresh injection of energetic particles. However, classical doubles do not normally evolve into FR Is since the two types of source are associated with different galaxies: twin jet and fat double sources are associated with brighter galaxies. On the other hand, the disturbed morphology of the Fornax A optical counterpart, NGC 1316, may be due to low-mass gas-rich mergers. In this case, the jet may spray over a wide angle over time.

Hercules A (3C 348) is a remarkable radio galaxy consisting of jets, rings, bulbs and bridges (Fig. 4.13) (Dreher & Feigelson, 1984). Although a relaxed double, it possesses an unusual jet-dominated morphology. It is classified as an FR 1.5. It is one of the most luminous radio sources in terms of both apparent and intrinsic brightness. The fourth brightest extragalactic source in the sky at low radio frequencies, its total power output is nearly as great as that of Cygnus A. It is identified with a very elongated cD galaxy (e.g. Sadum & Hayes, 1993) at the centre of a poor, faint cluster. This galaxy contains interlocking rings of extinction aligned near the radio axis (Baum *et al.*, 1996). The radio lobes are quite symmetrical and also contain ring-like/helical features in both lobes which form an almost symmetrical sequence. This suggests successive ejections from the active nucleus (Gizani & Leahy, 2003). However, no compact hot spots are detected.

4.2   Detailed structure and multiwavelength features

4.2.1   X-ray cavities and relics

The discovery of cavities in the intracluster medium with ROSAT (Boehringer *et al.*, 1993) and then the Chandra X-ray Observatory (Fabian *et al.*, 2000) has been an unexpected advance in our understanding of extragalactic jets. The X-ray cavities are directly associated with the extended lobes of radio galaxies residing in rich clusters. They are much rounder than the radio lobes and clearly have been inflated, as the leading bow shock associated with the jet has advanced through the ambient gas.

For Cygnus A, the gas emitting the bremsstrahlung X-ray emission is distributed within a prolate ellipsoidal shape out to about 110 kpc from the centre of the radio galaxy (Fig. 4.14), beyond which the distribution is spherical out to at least 700 kpc (Smith *et al.*, 2002). The structure is consistent with a scenario in which the intracluster gas has been swept up and compressed by an extremely broad jet cocoon – the cavity. To inflate such a cavity requires the injection of relativistic material from the radio jets via the hot spots.

There are two types of cavity. Some appear to be related to relic radio galaxies, a class of old sources in which the jets are believed to have shut down and the radio lobes are now coasting out like bubbles rising buoyantly and rapidly fading. Other cavities are associated with radio galaxies in which the jets are persistent.

There is evidence that the lobes of some radio galaxies are currently expanding very slowly, or possibly even stationary (McNamara *et al.*, 2000). This can be explained if the jets are extremely light so that they provide little thrust into the ambient gas. Furthermore, light jets should feed into very wide cocoons of radio-emitting plasma (Section 9.5), consistent with the bubble shape of the cavities. However, the energy contained in the X-ray cavity is known from the density and temperature required to account for the bremsstrahlung emission. The work expended to inflate the radio lobes can be simply calculated through

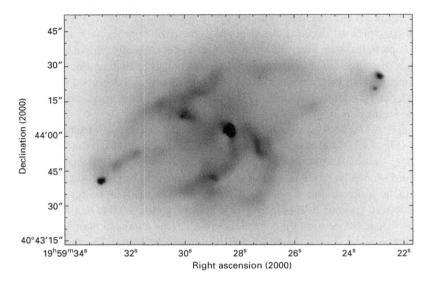

Fig. 4.14 The Cygnus A radio galaxy as observed in X-rays with the Chandra X-ray Observatory. Note the compact X-ray hot spots which correspond to the radio hot spots prominent in Fig. 1.2. The image, covering emission in the range 0.75 – 8 keV, has been processed to enhance the filamentary structure (Smith *et al.*, 2002). Reproduced by permission of the AAS.

the product $p\,dV$ where p is the pressure and dV is the volume of the cavity (e.g. De Young, 2006). Estimating the age of the radio galaxy, the power required to inflate the lobes far exceeds all other estimates of the amount of energy needed to supply the observed synchrotron radio emission. An extreme example of this is associated with MS0735.6+7421, where the average power required is 2×10^{46} erg s^{-1}, whereas the radio emission corresponds to a weak source, some 10^5 lower in power (McNamara *et al.*, 2005). The total energy released in the radio outburst must approach 10^{62} erg, sufficient to support the dense intracluster medium. While the jets must be the carrier of the energy, the source of the gas remains unclear. It could be resolved if the jets are dominated by a cold and relatively dense proton population which is hidden from our observations (De Young, 2006).

4.2.2   Hot spots

Hot spots are patches of enhanced radio emission embedded within radio lobes, usually located at the end of the lobes of powerful radio sources. These remote bright regions correspond to the 'working surfaces' of supersonic jets in which the bulk kinetic energy is abruptly dissipated within decelerating shock waves. Relativistic electrons are efficiently accelerated at the reverse shock front that forms in the jet due to the impact with the ambient medium. A second forward shock propagates into the ambient medium at a much slower speed but, nevertheless, has been detected in Cygnus A as a bow shock through a steep gradient in rotation measure (Carilli *et al.*, 1988).

At low resolution, hot spots in classical doubles such as Cygnus A were found to be simple features (Smith *et al.*, 1985). Two sub-spots were often found, a compact 'primary' hotspot and, in some cases, a diffuse 'secondary' hot spot. High-resolution images exhibit

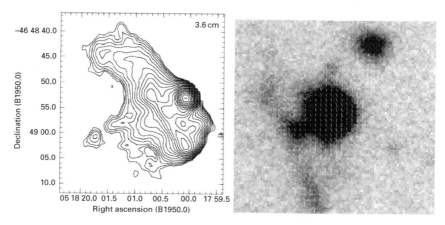

Fig. 4.15 The Pictor A hot spot in the western lobe. Left panel: the radio emission at 8.4 GHz. Note the bar-shaped region (the filament) just to the east of the compact hotspot. Right panel: optical Very Large Telescope image at 450 nm with polarisation vectors overlaid. Pictor A at a redshift of $z = 0.0342$ is an FR II radio galaxy. The images are not to scale (Saxton *et al.*, 2002).

a wide range of complex hot spot structures. These results contrasted with those of earlier studies of the more luminous objects (Leahy *et al.*, 1997; Saxton *et al.*, 2002).

The shock interpretation is consistent with the synchrotron spectral energy distribution between the radio and optical regime within hot spots as well as their radio morphology and polarisation properties. As the working surface advances, the relativistic particles are left behind, forming the extended radio lobes. The resulting synchrotron spectrum is expected to be steep at high frequencies due to the severe radiative losses that electrons experience as they escape the vicinity of the shock front (Meisenheimer *et al.*, 1989). Hence, optical observations of hot spots are rare. However, at even higher frequencies, the search for counterparts to the radio hot spots has been more successful. Many hotspots are detected in X-rays (Georganopoulos & Kazanas, 2003b) including the Cygnus A hot spots (Fig. 4.14) and the Pictor A western hot spot shown in Fig. 4.15. The radiation mechanism is not established, with synchrotron self-Compton, an additional synchrotron component or inverse Compton from the decelerating jet as proposed possibilities. For the specific case of the Cygnus A hot spots, the former mechanism is strongly favoured (Wilson *et al.*, 2000).

4.2.3 Optical and X-ray jets

The Pictor A jet is now one of many observed in X-rays with the Chandra X-Ray Observatory (Fig. 4.16). In the radio-loud quasar PKS 1127-145, the X-ray jet and radio jet intensity profiles are strikingly different (Siemiginowska *et al.*, 2007). The X-ray emission is strongest in the inner jet region and gradually decreases with distance from the core. The radio emission peaks at two outer knots. Given the issues with all one-component models, Siemiginowska *et al.* (2007) propose that the X-ray emission is from a jet spine while the bulk of the radio emission originates from a surrounding jet sheath.

X-ray jets are also observed in a number of FR I radio galaxies such as M 87 and 3C 66B (Hardcastle *et al.*, 2001). The Centaurus A X-ray jet consists of a low surface brightness

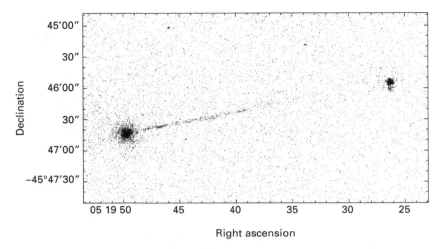

Fig. 4.16 The Pictor A X-ray jet, stemming from the core and directed exactly towards the distant western hot spot. The detected jet is up to 15 times brighter in X-rays than any counterjet, consistent with relativistic boosting. Image acquired by the Chandra X-ray Observatory in the 0.3 – 10 keV band (Wilson *et al.*, 2001). Reproduced by permission of the AAS.

component extending continuously from within 60 pc of the nucleus into the northeast radio lobe (Kraft *et al.*, 2002). The X-ray and radio morphologies of the inner jet are once again different but this time on the arcsecond level. Such offsets can be explained in terms of a combination of particle diffusion and energy loss.

In addition, bright optical knots are offset from X-ray knots in a number of jets including Centaurus A, 3C 31, 3C 66B, 3C 273, M 87 and PKS 1127-145 (Bai & Lee, 2003; Harris & Krawczynski, 2006). This is consistent with time lags introduced by synchrotron losses, suggesting that the large-scale optical and X-ray jets in these sources are attributable to synchrotron emission (Bai & Lee, 2003).

The Fermi Gamma-ray Space Telescope detected the radio galaxy Centaurus A in γ-rays (Fermi-Lat Collaboration *et al.*, 2010). The γ-rays originate from the giant radio lobes, well separated from the central active nucleus and provide constraints on the magnetic field and particle energy content in radio galaxy lobes. The emission from the lobes is interpreted as inverse Compton scattering of cosmic microwave background photons. An additional contribution at higher energies comes from the infrared-to-optical extragalactic background light. Gamma rays have also been detected from other radio galaxies including head-tail sources and is usually attributed to the base of the jet in the nuclear regions.

4.3   Host galaxies and triggering jets

The host of a radio galaxy is an elliptical galaxy (or a distorted version such as a merger remnant), with rare exceptions (Ledlow *et al.*, 2001). The elliptical radio galaxies are relatively larger than normal elliptical galaxies of the same absolute magnitude (Zirbel, 1996). It appears that powerful FR Is prefer their host galaxy that to be the dominant central galaxy of a relaxed group or clusters, while FR IIs are more likely to be associated with violent galaxy encounters. In either case, it could thus be interpreted that jets are only driven from the central engine and massive black holes in ellipticals. However, there is strong evidence for parsec- and kiloparsec-scale jets in spiral galaxies as well as small radio lobes (Baum *et al.*, 1993). Hence, it could be that nuclear jets are a common feature in Seyferts but the fragile jets are easily disrupted by the dense, rich interstellar medium of the spiral.

Many host galaxies of FR II radio galaxies at high redshift $z > 0.6$ appear elongated in the optical (McCarthy *et al.*, 1987). Intriguingly, the major axis is often aligned with the radio axis. Some galaxies were dissected into two colour components, an underlying red elliptical and a very blue elongated structure. Only the blue structure is aligned with the radio axis (Rigler *et al.*, 1992), and then this 'alignment effect' is only a property of cosmologically distant radio galaxies. This strongly suggests that the radio jets are not forced to propagate along the major axis but, instead, the jets generate the blue optical structure. In this scenario, the radio lobes heat and compress the gas-rich interstellar medium associated with these distant sources and so trigger large-scale star formation. Thus the formation and evolution of ellipticals associated with powerful radio galaxies and radio-quiet giant ellipticals may differ substantially as noted above. Further observations have supported jet-triggered starbursts as the origin of the alignment effect. Best *et al.* (1997) found that the jet from 3C 34 at a redshift of 0.69 produced intense star formation at a rate of $100 \, M_\odot \, \mathrm{yr}^{-1}$ on encountering a cloud in the cluster at a distance of 120 kpc from the nucleus.

4.4   Summary

Most galactic nuclei do not harbour measurable jets, or they possess jets with insufficient thrust to propagate out of the host galaxy. It is estimated that only about 10% of active galactic nuclei drive highly luminous radio jets, of which one in six is a classic double FR II radio galaxy (de Vries *et al.*, 2006).

The subject of extragalactic jets, even on just the large scale, has grown and diversified enormously. This observational review has therefore been limited to the essential properties associated with the jets themselves although we have touched upon the consequences. The dynamical and physical tools at hand appear to be sufficient but, despite all our efforts, fundamental questions remain open. On the large scale, we have a collection of snapshots to piece together. However, in the following chapter, we will approach the launch pad itself where the jet frames can be made into movies.

5

Jets in galactic nuclei

Single jets are seen to protrude from quasars as well as the cores of powerful radio galaxies. There is no physical principle which precludes intrinsically asymmetric systems, although a collapsed spinning object might be expected to possess a high degree of mirror symmetry. On the other hand, a system of merging binary black holes would certainly not. Nevertheless, the appearance of jets as one-sided led to the hypothesis that intrinsically symmetric twin jets contain radiating plasma that moves at relativistic speeds. The flux of one jet is Doppler-boosted for an observer lying within an emission cone centred on the jet direction, while the flux of the receding jet is reduced.

The discovery of knots of synchrotron emission seen to move in the sky away from quasar cores at speeds exceeding the speed of light turned into convincing evidence for the relativistic interpretation. The so-called superluminal motion was recorded in many quasar jets in the 1970s. The radio galaxy 3C 120 ($z = 0.033$) was observed at two epochs which suggested a speed of two to three times the speed of light (Shaffer $et\ al.$, 1972). The quasar 3C 345 was observed at four epochs in 1974 and 1975. The apparent transverse speed was reported as eight times the speed of light (Cohen $et\ al.$, 1976). As a result, an interpretation described in this chapter based on highly relativistic jet flows at a small angle to the line of sight became accepted.

It was further suggested that the variable radio emission from active galactic nuclei and quasars originates within the same collimated relativistic jet and that these also supply the extended radio galaxies with mass, momentum and energy (Blandford & Konigl, 1979b). The jet motions are observed in the form of bright *knots* or blobs of emission which are associated with velocity disturbances that steepen to form travelling shock waves.

We can naturally expect that in a small fraction of active galaxies or a tiny fraction of all galaxies, one of the jets will happen to be pointing more or less directly towards us. Combined with a highly relativistic flow speed, this leads us to expect a class of observed jets with extreme properties including apparent motions faster than light, apparent luminosities many orders of magnitude larger than the entire harbouring galaxy and variability on typical timescales of hours to years. The violent variability that is indeed observed across the electromagnetic spectrum has led to the general designation of *blazars* for this class of object.

The power generator of a quasar or blazar is contained within a radio *core*, a compact flat-spectrum source of radio emission. With a size of under 0.1", it requires radio interferometry techniques (very long baseline interferometry – VLBI) to be resolved. Cores are then observed to consist of components on milliarcsecond scales. These components are often aligned or show motions along a single direction, and are therefore related to the jet phenomena. With a growing ease of observation in well-concerted multiwavelength campaigns across continents, a detailed picture has emerged. We explore their remarkable properties in this chapter.

The true source of the jet was suspected to lie close to a specific flat-spectrum core on the milliarcsecond scale. This flat-spectrum core is stationary and was associated with the central black hole. However, it may well be associated with the detectable base of an optically thick jet or even a quite remote standing shock or bright knot well downstream. The other components move away at apparent superluminal speeds generating flat-spectrum and steep-spectrum wavelength ranges separated by a spectral break at the peak flux or turnover frequency. The knots' properties suggest that they expand as they propagate down the jet.

The reason why some active galactic nuclei present powerful radio jets and powerful radio lobes is a matter of debate. According to the standard model, a supermassive black hole with a mass between 10^6 and 10^9 $M_\odot$ resides in the nucleus of the active galaxy. It is powered by an accretion disc surrounded by a torus consisting of gas and dust. In about 10% of these AGNs, a radio galaxy is detected on extended scales. While there is mounting evidence for the existence of supermassive black holes at the centre of all AGNs (McLure & Dunlop, 2001), and even at the nuclei of non-active galaxies (Kormendy & Richstone, 1995; Macchetto, 1999), the source of the fuel for the jets is hotly debated. It has been argued that the presence or not of intense radio emission might be attributed to the rotation energy of the black hole (Wilson & Colbert, 1995), to the direction of the rotation (Garofalo *et al.*, 2010) or to its total mass and the efficiency of accretion (Dunlop *et al.*, 2003). A few case studies may elucidate the phenomena.

5.1   Individual blazar jets

5.1.1   3C 279

Despite the high redshift of 0.536, the blazar 3C 279 has occupied centre stage of multiwavelength observing campaigns. The coverage of the entire spectrum has made obsolete the alternative designations as a flat-spectrum radio quasar (FSRQ) and an optically violent variable (OVV), which no longer usefully encompass the remarkable variety of its behaviour. The black hole is estimated to have a mass of $M \sim 3\text{–}5 \times 10^8 M_\odot$. This yields the gravitational or Schwarzschild radius,

$$R_S = \frac{2GM}{c^2} \sim 1 \times 10^{-5} \frac{M}{10^8 M_\odot} \text{parsec}, \tag{5.1}$$

of 0.0004 pc. Thus, the event horizon of a non-spinning object is tiny in comparison with the resolvable size of jets with intercontinental radio interferometry (parsecs).

In 1971, radio images of both of the quasars 3C 279 and 3C 273 were reported to display rapid changes in their fine-scale structure (Whitney *et al.*, 1971). For 3C 279 this indicated a differential proper motion between components corresponding to an apparent speed of about ten times that of light. With later observations during the years 1991 to 1997 as well as a revised distance, apparent speeds measured for six superluminal components range from $4.8c$ to $7.5c$ (Wehrle *et al.*, 2001). On this parsec scale, as shown in Fig. 5.1, 3C 279 is quite complex. It consists of a bright, compact core, jet component knots and an inner jet that extends from the core to a stationary component one milliarcsecond (1 mas) from the core. Subsequent changes in the apparent speed near the core from $5c$ to $17c$ were recorded with

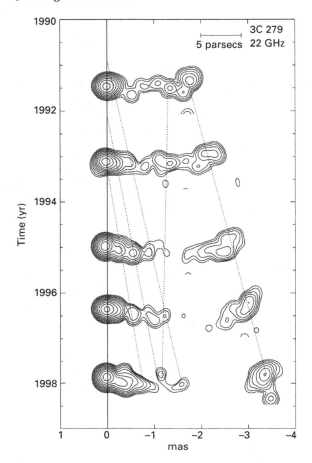

Fig. 5.1 A time-series of a selection of 22 GHz VLBI images of 3C 279 from epochs between 1991 and 1997 taken from Wehrle *et al.* (2001). The images have been restored and rotated. The solid line indicates the position of the presumed stationary core. The dotted lines represent the best fits to the model-fit Gaussian positions with time. The days of observation were 1991 June 24, 1993 February 17, 1994 September 21, 1996 May 13, and 1997 November 16. Reproduced by permission of the AAS.

the projected position angle of the jet skewing by about 20°. The skew could be brought about by a small change in viewing angle to the line of sight from ∼ 0.5° to ∼ 1.6° (Jorstad *et al.*, 2004). In comparison, the intrinsic jet opening angle is constrained to be smaller than about half a degree.

As well as the spatial changes, 3C 279 is also strongly variable in flux at all frequencies and a range of time scales (e.g. Chatterjee *et al.*, 2008). It varies dramatically from intense outbursts that may continue for one year to micro-variability on the timescale of hours. The variability profile of the flares is consistent with the optical emission of 3C 279 being synchrotron emission produced in the strong magnetic field of the relativistic jet. Furthermore, high-resolution radio maps show that the ejection of components from the core is synchronised with the flare activity.

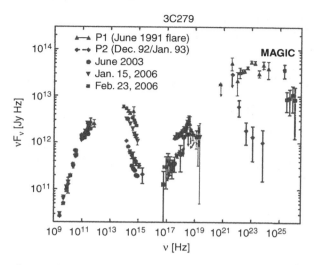

Fig. 5.2 The SED for 3C 279 as presented by Böttcher *et al.* (2009). Note the wide range of frequencies and the peaks in the infrared and beyond the hard X-ray regime. Reproduced by permission of the AAS.

The spectral energy distribution (SED) shown in Fig. 5.2 has two broad bumps with the first peak occurring in the far-infrared (10^{13} Hz) and the second one extending into the MeV–GeV energy range. In fact, 3C 279 was the first blazar discovered to emit at γ-ray wavelengths (by the EGRET on the Compton Gamma-Ray Observatory) and has now also been detected at photon energies above 100 GeV by MAGIC (MAGIC Collaboration *et al.*, 2008). The Fermi Gamma-ray Space Telescope discovered a γ-ray flare coincident with a dramatic smooth change in the angle of optical polarisation (Abdo *et al.*, 2010). This suggests that the optical and γ-rays are generated from the same jet location within which the magnetic field is uniform. Most significantly, this indicates that high-energy γ-rays, even when strongly flared, do *not* stem from the core or the vicinity of the black hole but originate from a location well down the parsec scale jet at 10^5 Schwarzschild radii.

The accepted interpretation of the first SED peak is that it is due to the synchrotron emission of highly relativistic electrons accelerated in a jet stemming from the nucleus. A plausible interpretation of the second peak is that it is generated by the inverse Compton (IC) emission of the same electrons interacting with low-energy photons. There are several candidates for the seed photons. They may be supplied

- by the jet itself (synchrotron self Compton, SSC),
- through infrared emission from warm dust associated with a surrounding molecular torus (external Compton, EC),
- by the clouds which constitute the broad emission line region (ECC) or
- directly from an accretion disc (EDC) and associated corona. The latter three are all termed external Compton, EC, processes.

In support of an IC model, variability in the infrared/optical is correlated with the GeV variations (Sambruna, 2007). The amplitude of the GeV variations went quadratically with the optical flux in early campaigns. This was interpreted as a variation of the electron density during flares in the SSC model (Ghisellini *et al.*, 1996) or a change of beaming factor in the EC model. During the 1996 campaign, however, the γ-rays varied more strongly than quadratically, and exactly (within one day) simultaneously with the X-rays (Wehrle *et al.*, 1998), favouring EC models.

To complete the AGN model, a search has to be made for the purported faint accretion disc which may supply the material and energy into the relativistic Doppler-boosted jet. Evidence for an underlying thermal component, consistent with emission from an accretion disc, has been noted by Pian *et al.* (1999) when 3C 279 was at a historical minimum at all wavelengths. This ultraviolet component suspected to originate from an accretion disc is detected in other blazars such BL Lacertae (Raiteri *et al.*, 2009) (the eponymous prototype of the blazars). It is detected only in low-activity phases when the beamed jet contribution of the synchrotron radiation is subdued. This could explain the smaller variability observed in the ultraviolet than in the optical, under the assumption that the disc emission remains constant over periods of less than a few years.

5.1.2   3C 273

The radio object 3C 273 is the brightest known quasar in the optical with a black hole mass estimated to be $1.4 \times 10^9 M_\odot$ (Labita *et al.*, 2006). Being relatively close at a redshift of $z = 0.158$ corresponding to 737 Mpc, a proper motion of 1.0 mas yr^{-1} would correspond to an apparent proper motion of $8.7 c$. This has made it a prime target for intensive monitoring with VLBI techniques for over 30 years (see Fig. 5.3). It has also

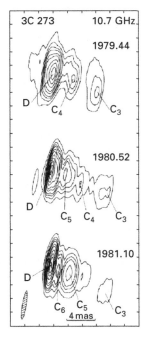

Fig. 5.3 The changing milliarcsecond jet of 3C 273 at three epochs between 1979 and 1981 in the radio at 10.65 GHz. The knots C3 and C4 were found to possess apparent angular speeds of 0.8 and 1.0 mas yr^{-1}, respectively, which now correspond to apparent speeds of $\sim 9c$. Taken from Unwin *et al.* (1985) (see also Pearson *et al.*, 1981). Reproduced by permission of the AAS.

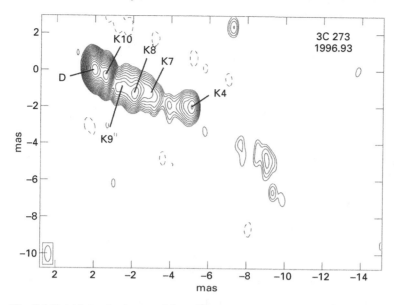

Fig. 5.4 Total intensity image of the milliarcsecond radio jet of 3C 273 at 22 GHz at the epoch 1996.93. Components with trajectories plotted in Fig. 5.5 are labelled. Reproduced from Homan *et al.* (2001). Reproduced by permission of the AAS.

been recorded for nearly 120 years in the visible with moderate variation in amplitude. The variability is at all frequencies, with amplitudes and timescales strongly depending on the energy and the different emission mechanisms involved.

The 3C 273 jet is one-sided with a small-scale radio jet on the parsec scale. This inner jet displays a stream of prominent knots as shown in Fig. 5.4. A new knot appears every few years from the flat-spectrum unresolved radio core and proceeds with a superluminal speed in the range $7c-11c$ (Figs. 5.4 and 5.5). Although components have been tracked which maintain constant speeds (Abraham *et al.*, 1996), different components can show a variety of accelerated motions.

The inner jet smoothly connects, while bending through $\sim 15°$, to a large-scale jet extending up to 60 kiloparsecs (Fig. 5.6). The extended radio jet is similar in detail to its optical counterpart (Fig. 5.7) with oblique structures in the radio jet corresponding to optical knots. Surrounding the knotty jet is a cocoon or sheath of radio emission.

A section of the inner kiloparsec of the jet has been resolved in a direction transverse to the flow (Savolainen *et al.*, 2006). It was found that there is a significant velocity gradient across the jet with the components near the southern edge being faster than the components to the north. This structure has been interpreted as a double helix consistent with a Kelvin–Helmholtz instability, and at least five different instability modes have been identified (Lobanov & Zensus, 2001). An interpretation as a light jet with a Lorentz factor of 2 and Mach number of 3.5 reproduces the major features. The low speed would suggest a two-component jet with a fast spine, in which shock waves form through the instabilities, surrounded by a slower sheath.

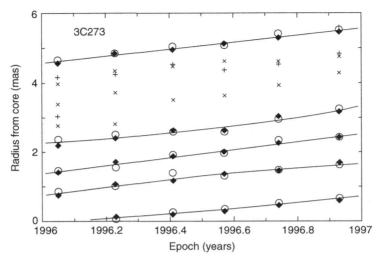

Fig. 5.5 The movement of radio knots in the milliarcsecond jet of 3C 273. Radial positions of the components labelled in Fig. 5.4 are plotted with filled diamonds at 15 GHz and large open circles at 22 GHz. Other symbols represent components with insufficient data to permit presentation of proper motions. Extracted from Homan *et al.* (2001). Reproduced by permission of the AAS.

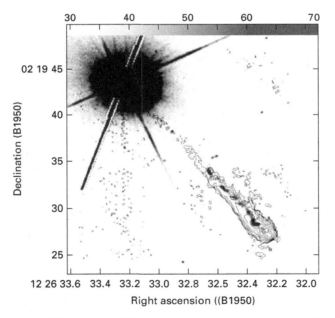

Fig. 5.6 The extended 3C 273 jet in radio (contours) and optical (greyscale) emission with resolutions of about 0.16″ and 0.1″, respectively. The optical image displays diffraction spikes which are used to locate the quasar centre. These images were taken with the Wide Field/Planetary Camera-2 on the Hubble Space Telescope (F606W wideband filter) and MERLIN at 18 cm, as presented by Bahcall *et al.* (1995). Reproduced by permission of the AAS.

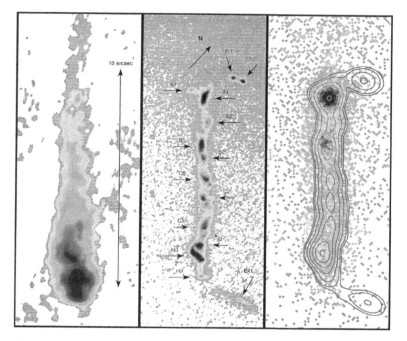

Fig. 5.7 The 3C 273 jet in three different bands with the quasar core to the top. Left: a MERLIN radio image at 1.65 GHz; middle: optical Hubble Space Telescope image in the wide 622 nm filter. Right: a Chandra X-ray image overlaid with a strongly smoother version of the HST image in order to match the X-ray resolution (Marshall *et al.*, 2001). Reproduced by permission of the AAS.

At present, neither the jet composition nor its velocity on the large scale is established. It has also been well studied, with Spitzer and Chandra demonstrating that it is indeed bright in the infrared and X-ray continua (see Fig. 5.7). In the infrared, as in the radio, the knots are brightest toward the far end of the jet. In contrast, the strongest X-ray emission comes from the first knot, with subsequent knots getting fainter along the jet (Marshall *et al.*, 2001). The origin of this X-ray emission is still controversial. Other problems remain unresolved, including the unusual morphology of the extended radio structure (with no lobes detected) and the exceptionally high brightness asymmetry (with a jet–counterjet luminosity of 10 000:1 (Conway & Davis, 1994).

The quasar spectrum is a superposition of components of comparable flux (Soldi *et al.*, 2008). At radio frequencies it appears as a core-dominated source of synchrotron radiation. The optical emission is dominated by the high-energy emission component of the jet, not by the radio synchrotron component, as had been assumed to date. The high-energy component, represented by a power law from the optical through X-ray, may be due to a second synchrotron component or to inverse Compton scattering of ambient photons.

The 3C 273 jet has been discussed in terms of jet precession, helical magnetic field, and growing plasma instabilities. In the precession model, the jet would have a bulk Lorentz factor of 10.8 and precess within a cone of half-opening angle 3.9° (Abraham & Romero, 1999). The Doppler factor should vary between 2.8 and 9.4 due to the changing viewing angle. On the other hand, the inclination of the nuclear jet is not expected to be smaller than 10°, because 3C 273 does not belong to the (regular) blazar class, owing to the presence of a thermal ultraviolet bump in its spectrum. The orientation of the large-scale jet differs from the orientation of the inner jet by about 20°, which is consistent with the apparent bend of the flow on the plane of the sky occurring at 10 milliarcseconds from the core. However, the high jet–counterjet asymmetry is not accounted for. Clearly, 3C 273 still keeps many secrets and remains high on the jet research agenda.

5.1.3   M 87

The giant elliptical galaxy M 87 is located at the centre of the Virgo cluster. Its fame rests on the non-thermal jet and the activity in the nucleus. At just 16 Mpc away, 1 milliarcsecond is equivalent to 0.08 pc and an angular speed of 1 mas yr^{-1} corresponds to 0.25 c. The nucleus is detected at all energies and is the only known non-blazar radio galaxy to emit very high-energy γ-rays. The black hole mass is $3 \times 10^8 \, M_\odot$ as measured through spectroscopy of rotating disc material within 3.5 pc of the black hole.

Although the M 87 jet dominates our attention, it is actually a part of a much larger radio galaxy, Virgo A. Low-frequency radio observations reveal a diffuse double-lobe structure extending to about 40 kpc as shown in Fig. 5.8. Two even larger bubbles of synchrotron emission towards the north-east and south-west appear to be inflated by this flow (Owen *et al.*, 2000). These bubbles are bright at the periphery and contain brighter wisps and filaments. It is morphologically classified as a type FR I radio galaxy (see Fig. 5.9) although its luminosity ($P(178 \text{ MHz}) = 1.0 \times 10^{25}$ W m^{-2}) is near the FR I/FR II border and it is best considered to be a transition or hybrid radio galaxy.

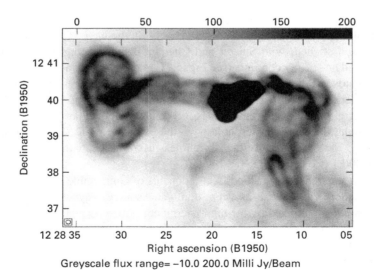

Greyscale flux range= −10.0 200.0 Milli Jy/Beam

Fig. 5.8 The main M 87 radio galaxy taken with the VLA at 90 cm, reproduced from Owen *et al.* (2000). The sharp drop in surface brightness between the inner lobes (containing the 2 kpc jet) and the outer structures is evident. Two bubbles extend to form a larger, almost circular halo around the entire radio galaxy. Credit: VLA, National Radio Astronomy Observatory/National Science Foundation.

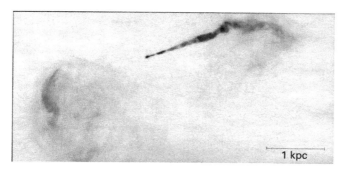

Fig. 5.9 The entire M 87 jet taken with the VLA in 1989. Credit: Archive of the National Radio Astronomy Observatory/National Science Foundation.

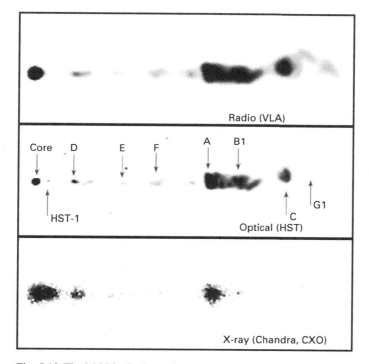

Fig. 5.10 The M 87 jet in the radio (top, VLA), optical (middle, HST) and X-rays (lower panel, Chandra). The image has been processed in order to emphasise the major emission regions. Credit: X-ray, NASA/CXC/MIT/H. Marshall *et al.*, 2001; Radio: F. Zhou, F. Owen (NRAO), J. Biretta (STScI); Optical: NASA/STScI/UMBC, Perlman *et al.* (2001).

On intermediate scales, M 87 displays a well-collimated, one-sided jet that is 2 kpc in length (Fig. 5.9). Its proximity and its misalignment from our line of sight permit detailed morphological studies at radio, optical, and X-ray energies as shown in Fig. 5.10. Knot A is clearly prominent in all bands. Apart from Knot A, the X-ray emission is strong in the inner jet while the outer jet is prominent in the radio.

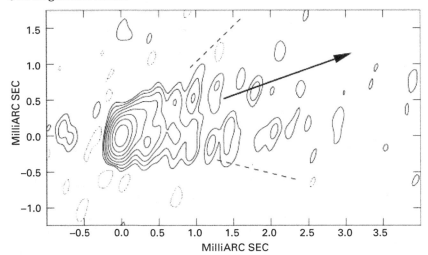

Fig. 5.11 The parsec-scale jet of M 87. A contour plot of the M 87 radio jet at 7 mm taken in 1999 with the VSOP+Global Array and presented by Biretta *et al.* (2002). The arrow points in the direction of the 20" jet while the dashed lines indicate the position angles of the limb-brightened structures within 1 milliarcsecond, as identified by Biretta *et al.* (2002).

On the parsec scale, the initial jet opening angle is approximately 60° on scales of about 0.04 pc, as shown in Fig. 5.11. The opening angle decreases rapidly thereafter until reaching 10° at a distance of 4 pc from the core (Biretta *et al.*, 2002). These observations suggest that the jet of M 87 is only gradually collimated, with the initial collimation taking place over 100 Schwarzschild radii. The jet maintains a constant orientation but contains much sub-structure on all scales, with many discrete knots spread along its entire length (Fig. 5.12).

In addition, the inner jet is seen to bifurcate, as is apparent on the VLBI image shown in Fig. 5.11, at distances exceeding about 5 milliarcseconds (0.4 pc) away from the core. The brighter jet limbs could be due to a physical splitting of the jet or due to the projection of the shell of a cylindrical jet (Kovalev *et al.*, 2007). The latter may be a consequence of a fast spine jet embedded within a slow sheath. In this configuration, the axial jet is beamed in a narrower cone away from the observer to yield a Doppler factor that varies across the jet. Note that the VLBI image also displays evidence for a counterjet extending toward the south-east as well as evidence for several stationary knots, both of which are consist with a spine–sheath model (Kovalev *et al.*, 2007).

Surprisingly, superluminal motion of radio features have been found in the M 87 jet at a site remote from the central black hole (Cheung *et al.*, 2007). One of the emission knots, HST-1d, is basically stationary to within 2 milliarcseconds (i.e., its speed is below $0.25c$) at a projected distance of 65 pc (860 milliarcseconds) from the core. However, this feature is the apparent point of origin of downstream superluminal knots: the knots appear to be ejected from and move away from knot HST-1d with complex speeds, typically between $0.4c$ to $4.3c$. The superluminal motions require that the angle to the line of sight is less than 30°, implying that HST-1 is intrinsically over 120 pc from the core. Superluminal motion

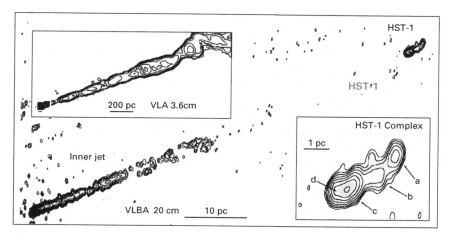

Fig. 5.12 The M 87 jet on various scales. The upper left inset displays the 2 kpc jet taken with the VLA at 3.6 cm image in December 2004. The main panel shows the inner 80 pc (VLBA image at 20 cm) with an enlargement of the optical HST-1 feature in the lower right inset. Image reproduced from Cheung *et al.* (2007). Reproduced by permission of the AAS.

has also been recorded throughout the inner $10''$ of the jet with speeds of up to $6c$. The structure of the inner jet is shown in Fig. 5.11. Gradual changes in the flux of individual knots have been followed (Biretta *et al.*, 1999).

The knot HST-1 is well isolated from the nucleus and the rest of the jet, so the strong flaring of higher-energy (X-ray, optical) emission detected by the Chandra X-Ray Observatory and the Hubble Space Telescope can be attributed to this site. During a monitoring programme of M 87, a rapid flare in the γ-rays was detected by the MAGIC telescope in early 2008 (Albert *et al.*, 2008). The γ-ray flux was found to be variable above 350 GeV on a timescale as short as one day at a high significance level of 5.6σ. The knot may well be responsible for a TeV flare observed by HESS around 2005. It has been suggested that HST-1 is a plausible site for the production of a dominant portion of the detected very high-energy TeV emission, including variations on the timescale of days, due to inevitable inverse Compton upscattering off the field of ambient photons by the electrons producing the flaring synchrotron optical-to-X-ray emission. From this, an emission region of size 0.002 pc is inferred. There is strong evidence for a frequency-dependent decrease in the X-ray synchrotron flux which would correspond to synchrotron cooling (Harris *et al.*, 2009). An average magnetic field strength of 0.6 mG for the HST-1 emission region is then derived, consistent with estimates for the equipartition field. Interestingly, the X-ray light curve also exhibited a quasi-periodic oscillation prior to the major flare in 2005.

This remarkable outburst activity observed in HST-1 is at odds with the common definition of active galactic nucleus variability usually linked to blazars, which originates in close proximity to the magnetosphere of the central black hole. In fact, the M 87 jet is not aligned with our line of sight and the HST-1 location corresponds to a distance of one million Schwarzschild radii from the supermassive black hole in the core of the galaxy. Thus, at least in the case of M 87, the observed hallmarks of blazar behaviour are not directly associated with the immediate vicinity of the black hole where the jet is launched, although, as shown by Acciari *et al.* (2010), the implications for each outburst are contrary.

5.2   Speed and Doppler boosting

The standard model for compact relativistic jets is presented here. There have been numerous alternative suggestions for the origin of the superluminal knots, some more contrived than others. Within the established jet context, one can consider the propagation of discrete blobs of plasma or turbulent 'plasmons' which expand as they move down the jet, with relativistic electrons being accelerated by the second-order Fermi process. Dynamical instabilities such as pinching instabilities or deflecting shock waves generated by jet grazing off ambient clumps could also be relevant. However, models which involve the development of shock waves resulting from central power surges have proved the most successful in terms of the interpretation of spectra and polarisation.

One-sided jets and superluminal motion within jets are simultaneous outcomes in the relativistic beaming interpretation of Blandford & Königl (1979b) which now provides the accepted interpretation. The basic results, which can all be derived through straightforward mathematical procedures, are presented here. The apparent or observed motion v_{ob} may far exceed the true motion, $v_k = \beta c$, when the emitting knot of plasma approaches us at nearly the speed of light along a small angle to the line of sight. Therefore, along our line of sight, the knot almost keeps pace with photons it emitted earlier. Hence the interval between arrival or receipt times is reduced relative to that of the rest frame of the knot. It is indeed straightforward to show that

$$v_{ob} = \frac{\beta c \sin \theta}{1 - \beta \cos \theta},$$ (5.2)

where θ is the angle between the velocity vector and the line of sight to the observer (Ginzburg & Syrovatskii, 1969). Here, the numerator, $\beta c \sin \theta$, represents the projected distance on the sky moved per unit time. The denominator, when multiplied by c, $1 - \beta \cos \theta$ is the difference in distance travelled towards the observer between the photon and the knot per unit time. Hence, when divided by c, this yields the difference in times of receipt of photons by the observer.

This is very neatly written in vector notation as

$$\frac{\mathbf{v}_{ob}}{c} \equiv \boldsymbol{\beta}_{ob} = \frac{\mathbf{n} \times (\boldsymbol{\beta} \times \mathbf{n})}{1 - \boldsymbol{\beta} \cdot \mathbf{n}},$$ (5.3)

where $\mathbf{n}$ is the direction toward the observer and $\boldsymbol{\beta}_{ob}$ is the velocity of the knot.

The observed speed depends on the angle of the motion to the line of sight. For a small angle θ, the apparent speed is small, while for $\theta = 90°$ the speed is less than c. The maximum apparent speed is found (on differentiating Eq. (5.3) and setting to zero) to occur at the angle $\theta = \sin^{-1}(1/\Gamma)$, where Γ is the knot Lorentz factor, $\Gamma = (1 - \beta^2)^{-1/2}$. This maximum speed is $\Gamma v_k = \Gamma \beta c$, which can clearly far exceed unity and approaches Γc for an ultra-relativistic knot.

The intrinsic acceleration of this knot is

$$\mathbf{g}_k = \frac{d(\Gamma \boldsymbol{\beta})}{dt} = \Gamma \frac{d\boldsymbol{\beta}}{dt} + \Gamma^3 \left(\boldsymbol{\beta} \cdot \frac{d\boldsymbol{\beta}}{dt} \right) \boldsymbol{\beta}.$$ (5.4)

However, the apparent or observed acceleration will be

$$\mathbf{g}_{ob} \equiv \frac{d\boldsymbol{\beta}_{ob}}{dt_{ob}} = \frac{(1 - \boldsymbol{\beta} \cdot \mathbf{n}) d\boldsymbol{\beta}/dt + (\mathbf{n} \cdot d\boldsymbol{\beta}/dt)(\boldsymbol{\beta} - \mathbf{n})}{(1 - \boldsymbol{\beta} \cdot \mathbf{n})^3},$$ (5.5)

where t_{ob} is the observer's time.

If the acceleration is radial, i.e. $\mathbf{g}_k \times \boldsymbol{\beta} = 0$, then $\mathbf{g}_{ob}$ is collinear with $\mathbf{v}_{ob}$ with magnitude

$$g_{ob} \equiv \frac{d\beta_{ob}}{dt_{ob}} = g_k \mathcal{D}^3 \sin\theta, \tag{5.6}$$

where $\mathcal{D}$ is the all-important Doppler factor defined as

$$\mathcal{D} = \frac{1}{\Gamma(1 - \beta\cos\theta)}. \tag{5.7}$$

Note that the maximum value of g_{ob} is $g_k \sin^2\theta$.

The Doppler factor is the standard formula in special relativity for the frequency ratio between observed and emitted photons, $\mathcal{D} = \nu_{ob}/\nu$. The luminosity ratio $\mathcal{L} \equiv L_{ob}/L_k$ for a moving knot of luminosity L_k can be shown to be $\mathcal{L} = \mathcal{D}^4$, where a factor of 2 in the index arises from relativistic aberration or beaming, one through the increase in photon energy and a further one for the time compression as the knot approaches. Alternatively (or equivalently), the intensity ratio $\mathcal{I} \equiv I_{ob}(\nu_{ob})/I_k(\nu) = \mathcal{D}^3$ since $I(\nu)/\nu^3$ is a Lorentz invariant.

Together with the shape of the spectrum, the Doppler factor also determines the increase in flux density, S_{ob}, over that emitted in the frame of the knot, S_k, at the observed frequency. On taking a power law with spectral index $\alpha = -d\log S_k/d\log\nu$, the flux density is

$$S_{ob}(\nu) = \mathcal{D}^{3+\alpha} S_k(\nu). \tag{5.8}$$

This result holds for optically thin or spherically symmetric optically thick knots. The Doppler factor is a maximum at $\theta = 0$ where $\mathcal{D} = \Gamma(1 + \beta)$.

For a knot undergoing linear acceleration, the beaming will at first increase the flux until $\beta = \cos\theta$, and with continued acceleration the flux then declines as the radiation is beamed into a narrower cone away from the observer's direction. On the other hand, the beaming pattern becomes wider for a jet which is radially decelerating. If observed at small angles, the observed spectrum is dominated by synchrotron emission from the base of flow, which may be unaffected by radiation losses. For those jets observed at larger angles, the downstream structure will dominate the spectrum (Georganopoulos & Kazanas, 2003a).

For a distant object, the redshift must also be accounted for. The total frequency shift is then given by

$$\nu_{ob} = \frac{\mathcal{D}}{1+z}\nu_k, \tag{5.9}$$

and the flux is rewritten as

$$S_{ob}(\nu) = \left(\frac{\mathcal{D}}{1+z}\right)^{3+\alpha} S_k(\nu). \tag{5.10}$$

A jet is less beamed than a blob. For relativistic motion along a continuous jet of fixed length which stems from a fixed source, each emitting particle can be considered as later replaced by another one emitting from the same position. Hence, the time compression factor reduces to just a time dilation factor of $1/\Gamma$ and the jet luminosity is $\mathcal{D}^3/\Gamma$ (e.g. Sikora *et al.*, 1997; Stawarz *et al.*, 2003). Relevant to observations are the measurement of flux density at a particular frequency. In this case,

$$S_{ob}(\nu) = \mathcal{D}^{2+\alpha} S_{jet}(\nu). \tag{5.11}$$

The jet 'sidedness' parameter thus predicted by relativistic beaming alone, assuming two equal and opposite flows of the same intrinsic length, is

$$\frac{S_{approach}}{S_{recede}} = \left(\frac{1 + \beta \cos\theta}{1 - \beta \cos\theta} \right)^{2+\alpha}. \tag{5.12}$$

This so-called Doppler favouritism is one of the major factors which are used to test unification schemes for galactic nuclei (Ghisellini *et al.*, 1993).

5.3   The class of blazar jets

The location of a flat-spectrum radio core is often assumed to be the site of the jet launch. Since jets expand even in recollimation zones (where the opening angle decreases although the radius increases with distance), the drop in density associated with the expansion implies that the highest densities are associated with the launch zone (although the local density maximum will also be associated with shock waves). Hence, a radio knot exhibiting synchrotron self-absorption will most likely be associated with the point of emergence of the jet from the launch region as it transformed from the optically thick regime into the optically thin regime.

Alternatively, the core component could already be well downstream in the flow from the launch region. In this case, higher-frequency observations would penetrate deeper upstream, an effect which is sometimes but not always observed. In addition, many jets possess one or more stationary knots or low-pattern-speed knots (Lister *et al.*, 2009) some of which can be identified as cores. These are often interpreted as standing shock waves, produced through recollimation. Such standing shocks in the jet could be oblique or conical, so dissipating only a fraction of the jet's kinetic energy.

Although each blazar jet has its peculiarities, there are common properties. Jets consist of a string of radiating knots and the properties of the knots are often attributed as the properties of the jet. Almost all jets appear one-sided (Lister *et al.*, 2009), consistent with differential Doppler boosting of intrinsically symmetric twin-jet propagation. Without exception apart from the stationary knots, the jet flow appears to be outward with speeds defined by Lorentz factors that are a characteristic of each individual jet.

The motions are in general directed along the jet ridge line. However, jet position angle swings have been recorded, including an angular rate of up to $11° \, yr^{-1}$ within the inner 3 pc of a jet (Agudo *et al.*, 2007) and in the BL Lac Object B2 1308+326, where the initial ejection direction appears to change with time. In some cases, such as 3C 279 or the γ-ray quiet quasar B2 0738+313, the flow appears to follow a predetermined curved or twisted trajectory and may not reach its final orientation until it is as much as 1 kpc or more away from the ejection point. One possible origin for the swings in jet direction may lie in the effects of precessional motion caused by a binary supermassive black hole.

The apparent speeds of features within a specific jet exhibit some spread. Taking a set of well-studied jets, the complete distribution of Lorentz factors extends up to $\Gamma = 40$ (Kellermann *et al.*, 2007). A larger data set reveals that the dispersion in speeds in specific jets is roughly three times smaller than the overall dispersion of speeds among all jets (Lister *et al.*, 2009). This suggests that there is a characteristic flow pattern pertaining to each jet which can possibly be characterised by the fastest observed component speed. The observed distribution of maximum speeds is peaked at $\sim 10c$, with a tail that extends out to $\sim 50c$.

This requires a distribution of intrinsic Lorentz factors in the parent population that range up to ~ 50.

However, velocity changes (in speed, direction or both) have been detected in many jets for individual knots as well as over much longer dynamical times. Several jets have displayed component acceleration with radial distance from the core. Accelerated motion is often found close to the origin followed by decelerated motion further out (Homan *et al.*, 2001).

The BL Lac objects are a class of blazars which do not display the strong emission lines of quasars. Named after the prototype BL Lacertae, they are characterised by rapid and large-amplitude flux variability and high optical polarisation in addition to the featureless non-thermal continuum (weak emission lines are detectable when the source is in the low-flux state). In the context of relativistic jets, we could expect these objects to form a subset of blazars in which we are peering almost exactly down the axis of the jet. This should clearly show up in terms of the knot motions. However, the opposite is found as discussed below.

An appropriate comprehensive study of jet speeds has been undertaken, by Jorstad *et al.* (2005), who presented total and polarized intensity radio images at millimetre wavelengths of two radio galaxies, five BL Lac objects, and eight quasars on angular scales 0.1 milliarcseconds. First of all, the Lorentz factors of the jet knots in the blazars are uniformly distributed in the range from 5 to 40 with the majority of the quasar components having 16–18. The brightest knots in the quasars have the highest apparent speeds, while the slower-moving components are brightest in the BL Lac objects. Jorstad *et al.* (2005) found that the quasars have similar opening angles and marginally smaller viewing angles than the BL Lacs. As expected, the two radio galaxies have lower Lorentz factors and wider viewing angles than the blazars. Thus, opening angles and bulk-flow Lorentz factors are inversely proportional, as predicted by gas dynamical models.

In further support of this, Pushkarev *et al.* (2009) found that the apparent opening angles of γ-ray bright blazars are preferentially larger than those of γ-ray weak sources. As noted for the examples above, γ-ray emission is directly associated with blazar jets. This fits into the picture that the γ-ray bright jets are preferentially oriented at smaller angles to the line of sight, resulting in larger Doppler factors (Pushkarev *et al.*, 2009). The intrinsic jet opening angle and the Lorentz factor were again found to be inversely proportional, as predicted by the standard model of compact relativistic jets presented above.

The mean values of the bulk Lorentz factor and the angle to the line of sight derived for the superluminal sources are consistent with those required to unify BL Lac objects with Fanaroff–Riley Type I, and radio-loud quasars with Fanaroff–Riley Type II radio galaxies. The classes differ in many properties including their cosmological evolution, with quasars generally at higher redshift. The viewing angles and the Lorentz factors were recently determined for 53 sources (38 flat-spectrum radio quasars) and 15 BL Lacs) with available superluminal velocity by Fan *et al.* (2009). They found mean values of $\Gamma = 24.1 \pm 17.3$ and $\theta = 8.7° \pm 3.5°$ for the quasars, and $\Gamma = 13.5 \pm 11.2$ and $\theta = 12° \pm 7°$ for the BL Lacs. Hence the BL Lacs are less beamed than the quasars. However, unification schemes and the models for γ-ray jets produce fresh disagreements, discussed in Section 5.5.

5.4 Variability and temperature

The violent activity of blazars and their observed high brightness temperatures (Ulrich *et al.*, 1997) provide strong arguments for an interpretation in terms of bulk relativistic motion close to the line of sight. The Compton Catastrophe is a long-standing issue which initially led some to question the cosmological distances of quasars. The problem arises from considering the consequences of the continuous non-thermal emission from the radio into the infrared which is incoherent synchrotron radiation. In the synchrotron self-Compton (SSC) process, some of the synchrotron photons will be inverse Compton scattered to higher energies by the relativistic electrons. If the photon energy density in the emission region exceeds the energy density of the magnetic field, catastrophic radiation losses must occur (Kellermann & Pauliny-Toth, 1969). For some blazars, the synchrotron radiation density inferred from the observed radio power and angular size predicts SSC X-rays well in excess of the observed X-ray flux, the apparent contradiction being termed the *Compton catastrophe* (Hoyle, 1966). The maximum steady brightness temperature of a circular Gaussian observing beam is given by

$$T_B = 1.22 \times 10^{12} \frac{S(1+z)}{a^2 v^2} \text{ K,} \tag{5.13}$$

where S is the flux density of the Gaussian in janskys, a is the full width at half maximum of the Gaussian in milliarcseconds, and v is the observation frequency in GHz.

Variability in both flux density and spectral shape across the entire spectrum are a common property of blazars. Extremely rapid radio variability in some blazars requires extremely compact emitting regions. For certain sources, the variability timescales t_{var} are so short that the radio brightness temperature, based on an estimated source size of ct_{var}, is in clear conflict with this inverse Compton limit of $\sim 10^{12} K$. For example, the object S5 0716+71 was found to have a brightness temperature exceeding $10^{14} K$, in clear violation of the inverse Compton limit (Ostorero *et al.*, 2006).

As already discussed, the blazar phenomenon is an amalgamation of two classes of active nuclei. BL Lacertae objects possess weak or absent broad emission lines. Flat-spectrum radio quasars (FSRQs) possess strong and broad optical emission lines. Their spectral energy distributions are similar, being characterised by two broad humps as shown in Fig. 5.2. The low-frequency hump can peak anywhere in the range from the infrared to the ultraviolet, and is attributed to synchrotron emission from relativistic electrons in the jet. The high-energy hump extends up to GeV and TeV γ-rays.

The X-ray emission is often an extrapolation of the tail of the first synchrotron hump in BL Lacs, while in more powerful FSRQs the X-ray flux belongs to the high-energy hump produced by the inverse Compton process (Sambruna, 2007). This has been discussed in the context of a luminosity sequence. As the bolometric luminosity increases, the first synchrotron hump peaks at progressively lower frequencies, peaking in the infrared-optical for the FSRQs. On the other hand, in the lower-luminosity and line-less BL Lacs, the synchrotron peak shifts up to higher frequencies.

Similarly, the second hump in the SED also shifts its peak with luminosity. The second hump is due to inverse Compton scattering of ambient photons, either internal to the jet (synchrotron self-Compton, SSC) or external (external Compton, EC) as discussed for 3C 279 above. The 3C 279 data were fitted with a SCC+EC model by Giuliani *et al.* (2009). The emitting blob moves with a bulk Lorentz factor of 13 at an angle of 2° to the line of sight. The relativistic Doppler factor is then $\mathcal{D} = 21.5$. Giuliani *et al.* (2009) applied a double power-law distribution for the electron's energy density, with index $p_1 = -2.0$ for energies given by the electron Lorentz factor in the range $100 < \gamma_e < 600$ and $p_2 = -4.0$ for $600 < \gamma_e < 6.000$ with a density at the break of 30 cm^{-3}. The emitting blob has a

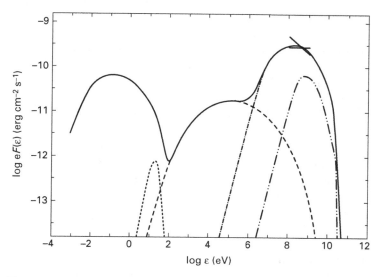

Fig. 5.13 The spectral energy distribution of 3C 279 from the AGILE-GRID observations, including simultaneous optical (REM) and X-ray (Swift) data. The dotted, dashed, dot-dashed and double-dot-dashed lines represent the contributions of the accretion disc black-body, the SSC, the external Compton scattering of disc radiation and the external Compton scattering of radiation from clouds, respectively (Giuliani *et al.*, 2009).

radius 2.5 10^{16} cm and magnetic field 1.8 G. The accretion disc luminosity assumed is 5×10^{45} erg s^{-1} with a broad line region reprocessing 10% of the illuminating continuum. The resulting spectral energy distribution is shown in Fig. 5.13, which fits the AGILE γ-ray, and simultaneous optical (REM) and X-ray (Swift) data.

The peak position in the SED may depend on the jet power. The more powerful jets are expected to have more efficient cooling in operation. It follows that the luminosity sequence for blazars can be interpreted as a manifestation of particle acceleration processes and cooling processes in the jets (Ghisellini *et al.*, 2002). In this picture, the relativistic electrons in the BL Lac jets suffer much less energy loss and continue to emit at the higher frequencies.

Relativistic beaming accounts naturally for the Compton catastrophe by introducing strong anisotropy and shortened timescales. Intrinsic brightness temperatures are raised by the Doppler factor to the observed values. A lower limit to the Doppler factor, which is necessary, can be estimated from the ratio of predicted to observed SSC flux. It is actually found that the Doppler factor has lower limits both larger and smaller than unity, depending on the particular AGN (Hovatta *et al.*, 2009). One complication is that the angular size is a function of observation frequency and so is to some extent arbitrary. Nevertheless, all quasars and a large fraction of BL Lacs are found to be significantly Doppler-boosted. In addition, there are some extreme sources with very high derived Lorentz factors of the order of a hundred. Alternatively, using variability timescales instead to infer Doppler factors from the T_B condition yields a Doppler factor exceeding unity for a number of blazars (Terasranta & Valtaoja, 1994).

Outbursts have often been observed to appear first at high frequencies with high amplitudes and then propagate to lower wavelengths with reduced amplitudes. This is probably consistent with the expansion of individual blobs as they propagate down the jet. Initially, the observed brightness temperature near the base of the jet may reach 5×10^{13} K, which is well in excess of the inverse Compton limit and corresponds to a large excess of particle energy over magnetic energy. However, more typically, the observed brightness temperaturese are $\sim 10^{12}$ K, closer to equipartition. The brightness temperature drops more abruptly with distance from the core in BL Lac objects than in the quasars and radio galaxies, perhaps owing to stronger magnetic fields in the former resulting in more severe synchrotron losses of the highest-energy electrons. However, the data interpretation is tentative. For example, the emitting jets are probably bent dynamic structures, and the changes in the viewing angles of the emitting regions are inevitable, with overwhelming consequences on the multiwavelength behaviour (Raiteri *et al.*, 2010).

5.5   The Lorentz factor crisis

An inconsistency has now arisen between two independent estimates of the jet Doppler factor. On the one hand, very high Lorentz factors are deduced from the teraelectronvolt γ-ray BL Lac objects based on the synchrotron self-Compton (SSC) models. On the other hand, a much lower Lorentz factor is favoured from the unification models between blazars and radio galaxies based on the statistics of beamed versus unbeamed objects.

The TeV γ-ray data mainly correspond to high-frequency-peaked BL Lac objects sources. Two of these blazars, PKS 2155-304 and Mrk 501, display variability on a remarkably short timescale of a few minutes, as observed by HESS (Namibia) and MAGIC (Canary Islands), respectively (Aharonian *et al.*, 2007; Albert *et al.*, 2007). This variability is shorter even than the hour-long timescale corresponding to the Schwarzschild radius of the black holes. In addition, the compact size of the region, if unbeamed, implies such a high number density of photons capable of pair production that they will not be able to escape the region (the γ-ray opacity problem described by Begelman *et al.* (2008)). The only answer to both these issues is to introduce bulk relativistic motion with a Lorentz factor as high as 100, invoked by Kusunose and Takahara (2008) to explain the variability and factors of $20-40$ needed to interpret the γ-ray MeV-GeV-TeV spectral energy distribution as obtained with the aid of the Fermi Gamma-Ray Space Telescope (Tavecchio *et al.*, 2010) with SSC single-zone models. These data have to be reconciled not only with the statistics of a unified model but also with the superluminal motions, if any. In fact, both methods suggest fairly modest bulk Lorentz factors for the parsec-scale radio jets of TeV BL Lacs objects of $\sim 3-4$ (Henri & Saugé, 2006; Piner *et al.*, 2008).

To resolve this crisis, we clearly need to dismiss the SSC one-zone model, although it is not clear how two regions with distinct Lorentz factors would be arranged and what underlying physics might be responsible. Clearly, with a decelerating jet, the radio knots lie downstream. The upstream energetic electrons from the fast base of the flow then Compton scatter the synchrotron seed photons, produced in the slow part of the flow, which are seen as relativistically beamed (Georganopoulos & Kazanas, 2003a). The opposite extreme is a stratified jet with an interacting spine–sheath or *needle-jet* structure (Ghiselnini *et al.*, 2005) in which one component is influenced by the beamed radiation produced by the other. This amplifies the inverse Compton emission of both components. Alternatively, Lyutikov and

Lister (2010) have suggested that the highly variable γ-ray emission is produced by the faster-moving leading edges of highly magnetised non-stationary ejections, while the radio data trace the slower-moving bulk flow.

There are numerous other effects in jets which may influence the structure, including opening angle effects (Gopal-Krishna *et al.*, 2004) and oblique shocks that result from the interaction of the jet with ambient matter. In the latter scenario, radio observations reflect the pattern speed, while the Lorentz factor inferred from TeV observations is associated with the speed of the fluid passing the structure. Such reconfinement shocks may occur on the parsec scale, generating the blazar emission by upscattering photons originating from the broad-line region and the molecular torus (Sikora *et al.*, 2009). The models will be further discussed in Section 9.3.5.

5.6 Polarisation

Strong and variable linear polarisation across all wavelengths has long been observed in many blazars. On the extended scales, jets can possess polarisations of several tens of per cent, which implies a uniform magnetic field within the observing beam. The cores more usually display a few per cent polarisation, suggesting a large number of field cells or a chaotic field structure.

The direction of polarisation is observed to rotate systematically, vary erratically or remain fixed. Swings of several degrees can occur on a nightly or yearly timescale. Swings through angles exceeding 180° have been followed in the core of the blazar OJ 287 (Homan *et al.*, 2002). Although we expect the magnetic field to be transverse to the electric polarisation vector in plasma producing optically thin synchrotron radiation, relativistic motion aberrates the angle, which then tends to align the polarisation vector with the jet direction and thus makes it difficult to deduce the intrinsic magnetic field direction. Acceleration within a relativistic jet may cause the polarisation vector to rotate through a large angle. Under some conditions, especially associated with flatter spectra, a swing in jet direction may cause the polarisation position angle to undergo abrupt 90° changes (Lyutikov *et al.*, 2005).

On the other hand, if the polarisation angle remains aligned with the jet direction even as it bends, the magnetic field structure must be dominated by an ordered transverse component (O'Sullivan & Gabuzda, 2009). In order for the jet polarisation to be oriented along the jet axis, the intrinsic toroidal magnetic field (in the frame of the jet) should be of the order of or stronger than the intrinsic poloidal field. Transverse shock waves would also have the effect of increasing the transverse component; the compression can initiate a swing through over 180° (Königl & Choudhuri, 1985b).

5.7 Summary

The entire field associated with active galactic nuclei is a large part of astrophysics. Jets are clearly so relevant that they are often synonymous with the AGN core activity itself, although something physical is hidden deeper. We have still not bridged the gap between the black hole and the milliarcsecond radio structure, a factor of 10 000, and there is no sign of that happening with emission from the radio to γ-rays deriving from components within the jet, often well downstream of the core which itself is part of the jet.

Relativistic effects dominate the blazar jets, making for fascinating extreme jet dynamics. However, there is no indication that we will have to revert to more exotic processes than those known and reviewed here. The determination of the physical structure and differential beaming transverse to the jet is imminent: resolving the velocity shear, magnetic field configuration and shock dynamics is feasible.

6

Jets from young stars and protostars

Directed flows of atoms and molecules are observed to stream away from young stars during their formation. Although the outflows are observationally prominent and suggest mass loss rather than gain, these early evolutionary stages are recognised as such by other signatures which indicate ongoing infall and mass accumulation. The infall continues from a surrounding envelope and through an accretion disc for the first few million years in the lifetime of a typical solar-mass or low-mass star, and probably on a shorter timescale for the formation of a massive star. At first, it appears paradoxical that infall should be so well signposted by outflow.

The accompanying outflow can take many forms besides that of a pair of highly supersonic antiparallel jets (Bally *et al.*, 1996). Often no jet or only one jet may be detected. Quite often, only a partly collimated outflow is observed in the form of two diffuse lobes, expanding in opposite directions: a so-called bipolar outflow. Then, jets are not necessary but there is a choice between jets and collimated winds to provide the thrust and supply the energy to drive the two large-scale reservoirs. In yet other cases, compact shocked knots are moving directly away from the young star. Often arc-shaped, they are interpreted to be bow shocks driven by jet-like flows (or jets which have dissolved into a chain of bullet-like projectiles; see Bachiller (1996)).

The jets can be seen as an inconsequential sideshow, to be given moderate attention only because they are spectacular. However, the more deeply we study these jets, the more we realise the intriguing role that the mechanism which is responsible for jets also plays in determining the nature of early stellar evolution. The jets, especially those from the earliest protostellar stage, carry away significant amounts of mass and energy from the active central region. Firstly, jets provide feedback into the molecular cloud which nurtures the driving star. Only a small part of these clouds will go on to collapse into clumps which, in turn, partially fragment into cores. Most of the core is probably returned into the cloud, and most of the cloud is also dispersed. It appears that the energy fed back from jets can be sufficient to make the difference between core dispersal and core collapse and so regulate the final mass of the system. Jets may do this by transferring some of the enormous gravitational energy of collapse back into the cloud in the form of supersonic turbulence.

Secondly, for stars to be able to form, the accreting material must lose almost all its angular momentum. In classical disc theory, the angular momentum is lost through outward transport within the disc via turbulent viscous forces and gravitational torques. In addition, the formation of secondary objects such as planets and low-mass stars provides another means of storing away the spin. However, it is purported that jets take over the mediation duties once the material approaches the inner magnetosphere of the star, extracting almost all the leftover angular momentum by expelling a small fraction of the inflowing mass. These regulation processes suggest that the physics which produces jets also influences the eventual mass of the forming protostar.

In addition, supersonic jets are important manifestations that herald the birth of young stars. When the formation process has reached its height, the interacting processes and components are hidden from our optical view by the dust in the cores and clumps. Despite this, by extending out of the core as collimated and ballistic flows that often stretch to parsecs, new stars can be found by tracing back along the trail it excites. In this manner, some of the deepest-embedded protostars have been discovered.

This chapter begins with a study of the older exposed T Tauri stars. With the Taurus star formation region at a distance of only 140 pc, their proximity offers a unique opportunity to investigate the accretion/ejection mechanisms at high angular resolution. This is fortunate since the jets, termed microjets, are indeed prominent on the smallest scales. We can now resolve the inner 50 AU, where we find that the jets are atomic and already high-speed. In fact, their speeds suggest that they were launched from much deeper within the gravitational potential well, within 0.1 AU of the young star.

At earlier stages, the jets are generally denser and more molecular. The propagating jet structure is often a series of knots described as a Herbig–Haro (HH) jet, with the HH prefix indicating that the emission is produced through shock excitation. The termination of the flow, or the end point of a particular outburst, often appears as a prominent Herbig–Haro object in the shape of a bow shock. Jets drive ambient molecular material forward and individual HH objects eject material sideways. The result is a reservoir of momentum which accumulates over time. Reservoirs are indeed found although it is never clear whether they have formed as a direct result of the jets or a surrounding wide-angle wind. The reservoir, however, is usually bipolar and this would then imply that the jets again come in antiparallel pairs.

Jets from young stellar objects (YSOs) are studied with a variety of techniques. Line emission permits speeds to be measured and ratios of line fluxes constrain the physical conditions. Multi-epoch imaging allows proper motions of jet knots to be measured. However, in YSO jets all these quantities may be related to internal shock waves or entrained gas rather than the original jet. Hence, modelling is often required to interface the shock properties with the shock-driving jet. This decoding is important to a giant Herbig–Haro flow in which the jet consists of a series of shocked knots that demarcate a meandering trail. Upon decoding, we can possibly infer some of the recent history relating to the accretion rate, orbital dynamics and system precession. Thus, the jet holds a fossil record of the activity of the driving source during star birth.

6.1   Optical jets

The discovery by Mundt and Fried (1983) of thin elongated optical nebulae directed radially away from several classical T Tauri stars confirmed that a form of jet was associated with young stars. T Tauri stars are optically revealed pre-main sequence stars which have already existed for roughly a million years. Although they have emerged from their native environment, stars of this classical variety are still actively accreting matter from circumstellar discs. The original stars found to harbour jets were DG Tau, DG Tau B, HH 30 and HL Tau, with a confirmation of the jet in IRS 5. All these jets were found to have typical lengths of $2-4 \times 10^{16}$ cm or ~ 2000 AU and small opening angles of 5–10°. Their collimation must take place at distances under 200 AU from the star.

Many discoveries have followed, especially through the Hubble Space Telescope (HST) and the application of adaptive optics to improve the spatial resolution to 0.1". The accreting T Tauri stars are now called Class II sources and the jets therefrom are almost exclusively atomic. The environments are too hot and too exposed to ultraviolet radiation to sustain molecules. The line fluxes are contained within radial distances of 1000 AU. After careful subtraction of the stellar continuum, microjet structures on scales under 100 AU are revealed in about 30% of the Class II sources. These are short, highly collimated optical jets with knots that display proper motions of a few hundred km s^{-1}.

The optical data sets are a rich source of information including multi-epoch direct imaging leading to proper motions. High-resolution spectroscopy reveals the line-of-sight motions. This emission is often blueshifted, as if the jet system were one-sided. In this case, the asymmetry is consistent with obscuration of redshifted emission from an outflow by an intervening circumstellar disc. However, star formation regions are complex systems with many interacting processes: all types of asymmetry including redshifted monopolar jets are also found (Henney *et al.*, 2002). The line profiles at the jet base are often multi-component (Hirth *et al.*, 1997). A high-velocity component and a low-velocity component are distinguished. The low-velocity component lies within a few tens of AU of the star, is blueshifted by only a few km s^{-1}, and appears strongest in lines with high critical densities. In contrast, the high-velocity component is more extended spatially, has velocities of hundreds of km s^{-1}, and a lower density (Hirth *et al.*, 1997).

The jets are observed in emission lines. Most significantly, physical conditions can be deduced from the forbidden emission line ratios. Going further, images of jets in a series of emission lines can be converted into synthetic images of the underlying physical parameters. This is accomplished by analysing the forbidden doublets [O I] with wavelengths of 630.0 nm and 636.3 nm, [S II] 671.6 nm and 673.1 nm, and [N II] 654.8 nm and 658.3 nm. The spectral images shown in Fig. 6.1 neatly display these lines relative to the Hα line at 656.3 nm. Although the jet Hα line is strong, it is contaminated by the prominent reflected component from the star. The technique presented by Bacciotti & Eislöffel (1999) yields the electron density n_e, hydrogen ionisation fraction x_e and electron temperature T_e from combinations of the above flux ratios of [S II], [O I] and [N II]. Dividing the electron density by the ionisation fraction yields the total hydrogen density n_H, and the product $T_e \times x_e$ yields an excitation image that is sensitive to hot ionised gas, as would be generated by a shock wave (Hartigan & Morse, 2007). The electron density up to about 2.5×10^4 cm^{-3} is sensitive to the [S II] ratio. The ionisation fraction derives mainly from the [N II]/[O I] ratio, and the temperature is partly indicated by the [O I]/[S II] ratio.

The flow of mass and energy within optical jets has been difficult to estimate in the past, given the issues in estimating extinction and the nature of photographic plates. Nevertheless, jet luminosities can now be derived from optically thin forbidden lines. In particular, HST spectra permit the conversion of flux directly into emitting gas mass. In this case, only the jet velocity is then required to obtain a mass outflow rate. The drawbacks remain the uncertainty in absolute calibrations and the possible strong and spatially variable extinction.

Alternatively, the mass outflow in the visible jet can be derived directly from the physical parameters derived from line ratios. The mass outflow rate is proportional to the derived density, radial velocity and cross-sectional area, which is estimated from the jet radius.

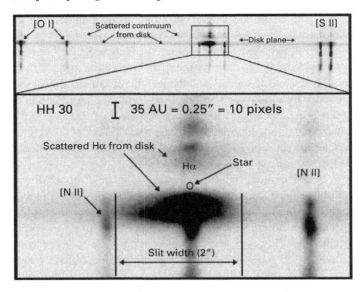

Fig. 6.1 Spectral images of the inner HH 30 system obtained with the Hubble Space Telescope as presented by Hartigan and Morse (2007). The use of a low-resolution grating with a 2" wide slit, much wider than the jet, generates an image for each emission line because the jet itself has a low radial velocity dispersion. The surfaces of the nearly edge-on opaque disc appear in reflected stellar light. The bright blueshifted inner section of the jet extends below the disc (to the northeast on the sky) in this spectral image. Credit: HST/STIS. Reproduced by permission of the AAS.

This method is not subject to reddening uncertainties since it is determined from line ratios. Differential reddening will only marginally alter the above line ratios.

Irradiated jets are detected in the optical where a jet has become exposed to an external ultraviolet radiation field from a neighbouring massive OB star. While the massive star dominates the atomic excitation and visibility (Reipurth *et al.*, 1998), the physical jets are produced by visible low-mass young stars which appear to have lost their parent cloud cores through photoevaporation and wind stripping. The jets are either one-sided or with the brighter jet pointing away from the irradiating star and about an order of magnitude brighter than the counterjet. Spectroscopy shows that the fainter counterjets are moving several times faster than the main jets. Thus the brightness asymmetry reflects an underlying kinematic asymmetry. These jets are often bent as if subject to a side wind (Bally & Reipurth, 2003).

6.1.1   *The HH 30 jets*

The HH 30 system has a fortunate geometry with a circumstellar disc that obscures the star but leaves the axis of the jet nearly in the plane of the sky. At a distance of approximately 140 pc, it is amenable to close inspection. As a consequence, a main jet and a counterjet have both been explored. This configuration was highlighted by the HST images which revealed that the surfaces of the opaque circumstellar disc appear as flared reflection nebula out to $\sim 400\,\mathrm{AU}$ (Burrows *et al.*, 1996). The models predict the disc inclination to be 82° with the brighter (northeast) side of the flow tilted toward the Earth. The reflection nebula varies significantly in brightness and in morphology, reflecting conditions present at the base of the accretion disc where material falls onto the star. The HST images show that this outer disc does not collimate the jet, which has emerged already collimated from the inner 20 AU. This is also consistent with the interpretation of jets driven magnetically from the very inner zones of accretion discs (Section 9.3.2). This is also consistent with the detailed disc structure determined by Guilloteau *et al.* (2008) from cold disc dust grains emitting in the millimetre continuum: the outer disc is truncated with an inner radius of ~ 40 AU. The system is a binary with a separation of ~ 15 AU, period 80 years and total mass $0.5\,M_\odot$ (Anglada *et al.*, 2007). Therefore, the outer disc is circumbinary and the optical jet originates from a region much smaller than 15 AU containing an inner compact disc.

The jet flow has been detected to within 20 AU of the star. New emission line knots emerge from this base exactly perpendicular to the disc plane on both sides of the disc. They are released every few years to replenish the collimated clumpy jets. The line widths in the jet are unresolved, so the flow must remain well aligned: there are no strong bow shocks to deflect the flow.

Individual knots exhibit distinct proper motions (Anglada *et al.*, 2007). As determined by Hartigan & Morse (2007), motions in the inner blueshifted jet range from 116 to 149 km s^{-1}, with uncertainties of just 4 km s^{-1}. Speeds in the redshifted counterjet range between 151 km s^{-1} and 229 km s^{-1}, significantly higher than they are in the main jet. On larger scales of up to $5'$, the motions were described as corresponding to a wiggling ballistic jet (Anglada *et al.*, 2007). This may be a consequence of the binary system with either jet precession due to tidal effects or orbital motion of the jet source around a primary as causes. Irregularities in the jet structure such as local variations in the width appear to develop near the base and are carried downstream with the flow. However, there is no clear correlation between the presence of a bright knot and the width of the jet.

This main jet has a resolved spatial width of FWHM ~ 14 AU at a distance of 20 AU from the star. The jet widens gradually to ~ 36 AU at 500 AU, with a constant opening half-angle of $2.6°$. This does not imply that the jet is produced directly from a disc of radius 14 AU. We expect that the opening angle of the flow is much wider nearer the source, as would be expected for magnetocentrifugal winds launched from an accretion disc.

HH 30 extends out to parsec scales from the source (Lopez *et al.*, 1996). The jet is not straight but curves, possibly as a result of precession. Proper motions again are quite disparate, covering a range $100 - 300$ km s^{-1} with considerable transverse motions (Anglada *et al.*, 2007). The jet not only has a change of direction of $14°$ in the plane of the sky (which is implied by the proper motion measurements) but is also curving towards us by an angle of $\sim 22°$.

The density in the jet is highest close to the source with a value of $2 - 3 \times 10^5$ cm^{-3} at 20 AU. As shown in Fig. 6.2, it subsequently declines in proportion to R_j^{-2}, in a manner similar to that of a radial conical flow at constant speed from a finite source region. The density also has a transverse gradient, being larger on the axis than along the edges by a factor of 2 (Hartigan & Morse, 2007). These values yield a mass outflow $\dot{M} \sim 2 \times 10^{-9}\,M_\odot$ yr^{-1} for each jet.

The excitation and ionisation exhibit a rapid rise followed by a gradual decline with distance from the star (Bacciotti *et al.*, 1999). Superimposed are high-excitation regions which possess sharp, almost linear boundaries that move outward with the flow. The regions of highest excitation are coincident with the side of the knot closest to the source.

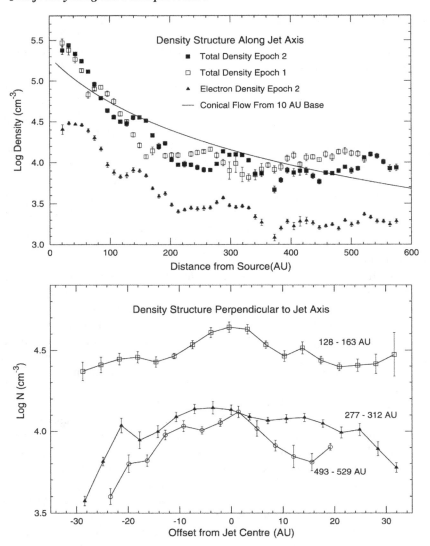

Fig. 6.2 The derived density distribution along the axis of the HH 30 jet (Hartigan & Morse, 2007). Both the electron density and the total density decline with distance from the source. The density fluctuations move along at the jet velocity between epochs. For comparison, the solid curve depicts the density fall in a conical flow of constant speed emerging from a base with a radius of 10 AU. Reproduced by permission of the AAS.

6.1.2   The RW Aur jet

The classical T Tauri star RW Aur is also a double system but with a wide separation. The primary, RW Aur A, is one of the optically brightest T Tauri stars. The secondary, RW Aur B, is located at a projected separation of 1.2". It is also suspected to be part of a multiple hierarchical system. A bipolar jet, with well-collimated blue- and redshifted lobes emanating from RW Aur A (HH 229), was found by Hirth *et al.* (1994). Further work traced the emission out to 15" in the redshifted lobe (Dougados *et al.*, 2000) and over 100" in the blueshifted beam (Mundt & Eislöffel, 1998).

Radial velocity measurements of several forbidden lines revealed an interesting asymmetry: the radial velocity of the blueshifted lobe reaches about -190 km s^{-1}, while a velocity of only $+100$ km s^{-1} is measured on the red side (Hirth *et al.*, 1994). Another unusual property of this jet is that for the first 10" the redshifted jet is brighter in the [S II] doublet than the blueshifted beam (Mundt & Eislöffel, 1998), whereas in general blueshifted jets are brighter. Beyond this distance, the redshifted RW Aur jet strongly drops in brightness, becoming much fainter than the blueshifted flow (Melnikov *et al.*, 2009). The redshifted jet contains more knots than the blueshifted one, suggesting more shock excitation.

The RW Aur jet is also knotty in the forbidden lines (Fig. 6.3). López-Martín *et al.* (2003) found that the mean proper motion of knots for the blue- and redshifted jets are 0.26" yr^{-1} and 0.16" yr^{-1}, respectively. This is proportional to their radial velocities, suggesting that these are material motions. The inferred inclination of the jets to the line of sight is $46° \pm 3°$.

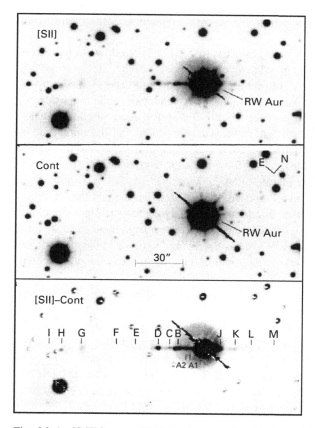

Fig. 6.3 An [S II] image of RW Aur (top panel), with the continuum image (middle) and with the continuum subtracted (bottom panel) (Mundt & Eislöffel, 1998). The ring-like emission is an artefact. The extended southeast jet contains at least 10 knots stretching over 100". The total spatial extent of the bipolar outflow is at least 145", corresponding to ~ 0.1 pc. Reproduced by permission of the AAS.

The jet knots have been detected in the infrared [Fe II] λ 1.644 μm line by Pyo *et al.* (2006). A comparison of knot positions suggests a good correlation with the optical knots taking into account the proper motions . At the same time, the HST data do not reveal any maxima at those positions in the optical. Instead, the forbidden emission flux steadily decreases from the position closest to the central source. Hartigan & Hillenbrand (2009) traced the flow in the [Fe II] line to a projected distance of only 10 AU from the source. The redshifted flow remains slower than its blueshifted counterpart. The radial velocities and the emission line widths are larger closer to the source on both sides of the jet. The line widths are 20–30% of the jet velocity on both sides of the flow, significantly larger than would be produced by a divergent constant velocity flow. The observed line widths could arise from a layered velocity structure in the jet or from magnetic waves.

The ionisation fraction, x_e, varies between 0.04 and 0.4, increasing within the first few arcseconds and then decreasing in both lobes. The ionisation fraction also shows asymmetries, while the electron density and temperature are similar in both jet lobes (Melnikov *et al.*, 2009). In the blueshifted lobe the average x_e is 0.23, while in the redshifted one it is 0.08. As a result, the redshifted jet is found to be 2 times denser. On the other hand the redshifted jet is about 2 times slower and this makes the mass outflow rates similar on the two sides.

Low average mass outflow rates along the first 2.1" of both flows (a region presumably not yet affected by interaction with the jet environment) of $2.6 \times 10^{-9}\, M_\odot\, \mathrm{yr}^{-1}$ for the red lobe and $2.0 \times 10^{-9}\, M_\odot\, \mathrm{yr}^{-1}$ for the blueshifted flow are derived. The mass outflow to mass accretion rate is 0.05 (Woitas *et al.*, 2002). At least part of the ionisation is produced and maintained locally at the internal shock fronts in RW Aur, where the peaks of n_e, x_e and T_e and the positions of the bright features are correlated. Furthermore, as in HH 30, the ionisation fraction tends to increase upstream of the knot, instead of at the knot itself.

Signatures consistent with rotation have been uncovered in both jets within the first 1.5" from the central source (about 200 AU) (Woitas *et al.*, 2005). Similar signatures of rotation with speeds $5 - 15$ km s^{-1} had been discovered in other jets from T Tauri stars(e.g. Coffey *et al.*, 2004). Such observations support the magnetocentrifugal launching scenario for jets which predicts a transport of angular momentum from the disc to the jet. Recently, however, the disc surrounding RW Aur has been found to rotate in the opposite sense to the bipolar jet (Cabrit *et al.*, 2006). An explanation in terms of precession rather than rotation is equally plausible.

6.1.3   The DG Tau jet

DG Tau hosts HH 158, one of the best-studied jets from a low-mass young star as a result of being the brightest jet in the optical. The blueshifted jet emits in two distinct velocity bins, a high-velocity component (HVC) and a low-velocity component (LVC) separated by ~ 100 km s^{-1} (Lavalley-Fouquet *et al.*, 2000; Coffey *et al.*, 2008). The [N II] emission is more collimated and traces high velocities, while [S II] is spatially extended and traces low velocities. These results were derived through remarkable high-resolution spectroscopy with the HST to obtain position velocity diagrams for spectroscopic slits placed transverse to the jet axis, as shown in Fig. 6.4. By using the line ratio methods, these diagrams are then converted into position–velocity diagrams for the physical parameters (Fig. 6.5).

The associated total density is as high as 10^6 cm^{-3} in the HVC and 5×10^5 cm^{-3} in the LVC. A typical electron temperature of 2.5×10^4 K was found for the HVC and 0.5×10^4 K

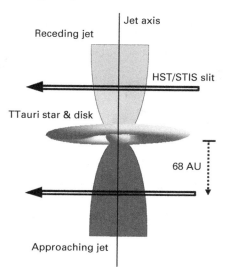

Fig. 6.4 The configuration for the Hubble Space Telescope STIS spectroscopic observations of DG Tau presented by Coffey *et al.* (2008). The arrow on the slit indicates the positive direction of the *y*-axis in the position–velocity diagrams. Reproduced by permission of the AAS.

for the LVC. Thus, at least in the case of DG Tau, the flow becomes gradually denser and hotter as it increases in velocity and collimation (Bacciotti *et al.*, 2000). The ionisation fraction is however, strangely, higher for the LVC than the HVC.

The relatively high brightness is a consequence of a relatively high rate of mass outflow, calculated from the above densities (Coffey *et al.*, 2008). The mass outflow rate is higher in the HVC at $4.1 \times 10^{-8}\ M_\odot\ \mathrm{yr}^{-1}$ compared with $2.6 \times 10^{-8}\ M_\odot\ \mathrm{yr}^{-1}$ in the LVC (Coffey *et al.*, 2008) although these values depend on where the separation is set. In comparison, the mass accretion rate for this system is much higher, at $2 \times 10^{-6}\ M_\odot\ \mathrm{yr}^{-1}$, yielding a ratio consistent with other values for other systems which are found to be $\dot{M}_{\mathrm{jet}}/\dot{M}_{\mathrm{acc}} \sim 0.01$–$0.07$.

The possible extraction of angular momentum through the DG Tau jet has also been investigated by Bacciotti *et al.* (2002). The velocity shifts are consistent with the southeastern side of the jet flow moving toward the observer faster than the corresponding northwestern side, assuming that the flow is axially symmetric. Interpreted as jet rotation, the jet rotates clockwise from the perspective of an observer looking down the jet towards the source with a toroidal velocity in the range 6–$15\ \mathrm{km\ s}^{-1}$. Employing the above estimates for the mass outflow rate, the rate of flow of angular momentum is $3.8 \times 10^{-5}\ M_\odot\ \mathrm{yr}^{-1}\ \mathrm{AU\ km\ s}^{-1}$ and is dominated by the HVC (Coffey *et al.*, 2008). This estimate is consistent with the rate of loss from an accretion disc as predicted by magnetocentrifugal jet-launching models, although the constraints are not strong and other causes of velocity gradients cannot be excluded.

The optical DG Tau microjet consists of a series of knots spread over 15". Given the proper motions of 0.15"–0.3" per year, the knots appear to be frequently ejected with tangential speeds ranging from $100\ \mathrm{km\ s}^{-1}$ to $230\ \mathrm{km\ s}^{-1}$ (Eislöffel & Mundt, 1998). Both the tangential and radial velocities change significantly from knot to knot with an inclination angle of the jet axis to the line of sight $\sim 32°$.

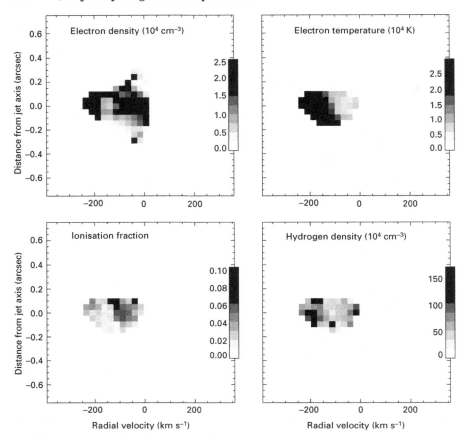

Fig. 6.5 The DG Tau jet. These position–velocity diagrams display the physical jet parameters as functions of distance from the jet axis and radial velocity. The panels exhibit the electron density (upper left), electron temperature (upper right), ionisation fraction (lower left) and hydrogen density (lower right) calculated for each pixel. Taken from Coffey *et al.* (2008). Reproduced by permission of the AAS.

X-ray emission from a thermal jet is only expected if fast shock waves are involved to heat the gas above 10^6 K. Although an X-ray emission region is associated with the base of the HH 154 protostellar jet (Bally *et al.*, 2003), the jet of DG Tau was the first to be confirmed to emit in X-rays (Güdel *et al.*, 2008). A prominent X-ray counterjet, has also been detected in the DG Tau system with excess absorption of the counterjet probably due to an intervening accretion disc. The X-ray jets are detected out to a distance of $\sim 5''$ from the star extending in the same direction as the optical jet but not so far out. The electron temperature is 3.4×10^6 K, much higher than the temperatures derived from the forbidden emission lines. The high temperature is also not consistent with the observed jet speed. This implies that the pressure in the hot gas contributes to jet expansion. This leads in turn to the suggestion that ohmic dissipation is involved. In this scenario, initially jet-aligned magnetic fields are wound up due to rotation, producing helical fields which drive currents (Güdel *et al.*, 2008).

6.1.4   *Optical jets: general results*

Optical jets are not just the original *microjets* of size 1000 AU but exhibit structure which suggests the jets remain in some form on scales of up to parsecs (Reipurth *et al.*, 1997a; McGroarty & Ray, 2004). For example, the DG Tau flow extends out to 17″ (1.1 parsecs) with HH 702 confirmed to be moving directly away from the young star (McGroarty *et al.*, 2007). Many optical jets, also called Herbig–Haro jets, have been investigated on these larger scales using ground-based telescopes. The hydrogen density of the contained knots is generally found to be between 600 and 6×10^4 cm^{-3}, one to three orders of magnitude lower than in the microjets. This is expected due to jet transverse expansion and sideways deflection at internal oblique and bow shocks.

There is also strong evidence that these outflows have drilled out of their parent molecular clouds, driving the shocked Herbig–Haro objects beyond the parsec scale. Typically, however, beyond the microjets, a series of spaced knots and/or bow shocks are detected instead of a smooth jet. This *fossil record* (Raga *et al.* 2002b) could imply that the sources have undergone FU Orionis-type outbursts on the timescale of 100–1000 years. In this regard, the dynamical timescale for flow along the jet is

$$t_j = 25.000 \left(\frac{\Delta D_k}{1 \text{ parsec}} \right) \left(\frac{v_j}{400 \text{ km s}^{-1}} \right)^{-1} \text{ yr}, \tag{6.1}$$

much shorter than the duration of the Class II T Tauri stage.

Mirror symmetry provides the strongest evidence that centrally generated disturbances generate the jet knots. Symmetrically located knots in twin-jet sources indicate a common dynamical time. Compression is caused by the fluctuations in ejection speed. This generates pressure waves which steepen as they propagate downstream. Within the pulses, compressed jet material accumulates to form dense knots called internal working surfaces. As the pulses move out along the jet, the strengths of the shocks decay and all the material is swept into the knots. In this manner, a continuous jet is transformed into a chain of bullets.

Even low-amplitude long-wavelength sinusoidal variations will steepen into shock waves as the waves propagate down the jet. Given the knot spacing, ΔD_k, and the jet speed v_j, we estimate the duration between pulses (or source outbursts) as $t_k = \Delta D_k / v_j$, which can be written

$$t_k = 1000 \left(\frac{\Delta D_k}{2000 \text{ AU}} \right) \left(\frac{v_j}{100 \text{ km s}^{-1}} \right)^{-1} \text{ yr}. \tag{6.2}$$

Hence, episodes of high accretion could lead to enhanced ejection events and so provide a natural explanation for jet knots. However, in order to operate, the ejection mechanism and the wave steepening process must together lead to jet velocity variations, rather than just to jet density or radius variations.

The supply of gas into the jet launch zone takes place through an accretion disc. The accretion may be smooth and regular in the early protostellar stage, since the disc evolution is expected to match the imposed infall rate. The inner disc is then sufficiently hot to be coupled to the magnetic field. However, in the Class II phase, the infall reaches the disc beyond ~ 10 AU and a *dead zone* occurs in the inner regions, where angular momentum transport through coupling to the magnetic field is inhibited (e.g. Zhu *et al.*, 2010). The disc then executes outbursts of accretion in FU Ori-like events.

Any model must satisfy the following general results.

- The collimation occurs on scales large compared to the radius of the young star. Jets are only weakly collimated on the scale of ~ 100 AU whereas the stellar radius is ~ 0.02 AU (Mundt *et al.*, 1991).
- Jets exhibit a monotonic decrease in opening angle with increasing distance from the source, finally reducing to typical full opening angles in the broad range of $0.5°$ to $5°$ (Ray *et al.*, 1996).
- The jets are knotty, with the knots typically displaying proper motion of 100–200 km s^{-1} (Eislöffel & Mundt, 1998).
- The line profiles are broad and often blue shifted to 100–400 km s^{-1}. The lower velocity is preferentially found around the edges and displays signs of acceleration with distance from the source (Pyo *et al.*, 2003).
- Jet speed tends to increase with the stellar luminosity.
- Variations in the jet speed are quite common.
- Asymmetries are observed in the radial velocity between the jet and counterjet.

The ionisation fraction is seen to rise with distance in the initial jet section in several sources (Coffey *et al.*, 2008). It reaches a plateau at $\sim 1'' - 2''$ or 100 AU from the source, and then falls slowly consistent with gradual recombination. The obvious interpretation is that the ions are produced close to the source through a series of discrete heating events. Re-ionisation events may occur downstream (e.g. Melnikov *et al.*, 2008). In HH 30, ionisation peaks follow the movement of the knots. Thus at least part of the ionisation is produced and maintained locally at the internal shock fronts. Furthermore, in HH 30 and RW Aur, the ionisation fraction tends to increase upstream of the knot, instead of precisely at the knot itself. In some jets in which re-ionisation episodes occur (HH 24C/E and HH 24G), the ionisation fraction abruptly rises and then falls smoothly downstream of the re-ionisation event (Bacciotti & Eislöffel, 1999).

The narrow opening angle of the HH 30 jet is not consistent with a freely expanding flow (Hartigan & Morse, 2007) for which the observed opening half-angle would correspond to a Mach number at the base of the jet of $M_j = 21.8$. The semi-opening angle consistent with thermal transverse expansion is $\tan \theta = c_s/v_j = 1/M_j$. Using a jet velocity of 130 km s^{-1}, we find a sound speed of $c_s = 6.0$ km s^{-1}, which corresponds to a temperature of only 2600 K for a ratio of specific heats of $5/3$ and mean molecular weight 1. The more likely sound speed in the HH 30 jet exceeds 10 km s^{-1} corresponding to 7260 K, which should occur for the observed low-excitation forbidden lines in a mostly neutral cooling zone of a shock.

The jet could be confined by an external pressure that is on the order of the thermal pressure, as discussed by (Hartigan & Morse, 2007). However, the thermal pressure of an ambient external medium may disrupt the jet through the growth of shear-driven Kelvin–Helmholtz instabilities at the interface between the two fluids. Instead, a toroidal magnetic field is invoked with a strength of perhaps 5 mG at 300 AU. Such strong magnetic fields are inevitable in centrifugally driven wind models. Nevertheless, although hoop stresses confine the jet, it is not clear how the toroidal field will itself be confined in the outer parts.

Finding an explanation for the observed velocity asymmetries such as in RW Aur remains an objective. As discussed by Melnikov *et al.* (2009), the mass outflow rate is actually about the same in both jets, which suggests that the origin of the differences in density, velocity and excitation may instead reside in the interaction with an asymmetric ambient medium. In a magnetocentrifugal launching scenario, a different magnetic field configuration on opposite sides of the disc could produce different jets by distorting the open field surfaces. In support, a number of young stars do show asymmetries in the magnetic field distribution on their surfaces. However, a connection between surface magnetic properties and the large-scale circumstellar magnetic configuration has yet to be investigated.

Estimates of the mass flux and angular momentum flux are fundamental aims in jet physics with which models of launch and accretion can be tested. Both present major challenges requiring measurements at the base of the jet. Further out, mass may have been loaded into the jet from the ambient medium and rotation speeds in the jet are difficult to measure due to jet expansion. Hence, HST data has provided the first opportunity to measure the jets from anywhere near their launch site. Coffey *et al.* (2008) find that for a single jet mass outflow rates and angular momentum outflow rates are in the range 4.0×10^{-9} to $6.7 \times 10^{-8} M_\odot \text{ yr}^{-1}$ and 1.1×10^{-6} to $1.3 \times 10^{-5} M_\odot \text{ yr}^{-1} \text{ AU km s}^{-1}$.

The mass flux through twin jets of radius r_j is $\dot{M}_{jet} = 2\pi r_j^2 \rho_j v_j$, assuming a circular cross-section across which the speed, v_j, and density, ρ_j, are constant. Class II optical jets typically possess high speeds and low densities. The overall result is a low mass loss rate:

$$\dot{M}_{jet} = 2.1 \times 10^{-8} \left(\frac{r_j}{100 \, \text{AU}} \right)^2 \left(\frac{n_j}{10^3 \, \text{cm}^{-3}} \right) \left(\frac{v_j}{400 \, \text{km s}^{-1}} \right) M_\odot \text{ yr}^{-1}. \quad (6.3)$$

The optical jets are quite powerful due to their speed. The power transported is

$$\dot{L}_{jet} = 0.13 \left(\frac{\dot{M}_{jet}}{10^{-8} \, M_\odot \text{ yr}^{-1}} \right) \left(\frac{v_j}{400 \, \text{km s}^{-1}} \right)^2 L_\odot. \quad (6.4)$$

These values are consistent with launch mechanisms involving accretion discs in their final stage. For example, an estimate for how the long-term accretion rate decreases with evolution time is

$$\frac{\dot{M}}{3 \times 10^{-8} \, M_\odot \text{ yr}^{-1}} \sim \left(\frac{t_{age}}{10^6 \, \text{yr}} \right)^{-1.5} \quad (6.5)$$

(Hartmann *et al.*, 1998). However, there are strong mass and environmental factors which probably control each evolutionary accretion track (Calvet *et al.*, 2004). Furthermore, in this stage, the gas still accretes but the dust grains may not be strongly coupled via collisions. The larger dust grains are left behind to form rocky objects which may be the seeds for planets.

6.2 Embedded protostellar jets

Jets of a more ballistic kind are detected in the stages before the young stars have appeared in the optical. Jets from these Class 0 and Class I protostars are found to be denser, of higher Mach number and more powerful. This activity is commensurate with the early high accretion rates and gravitational energy release during star formation. For the two example systems discussed here, HH 34 possesses a disc of mass 0.2 $M_\odot$ and HH 111 a disc of mass of 0.3 $M_\odot$ (Stapelfeldt & Scoville, 1993). These circumstellar discs are roughly ten times more massive than those surrounding T Tauri stars.

The jets drive large, slower-moving molecular lobes called bipolar outflows. Although, as reservoirs of expelled energy and momentum, the bipolar outflows are the equivalent of the lobes of radio galaxies, it is not clear that they are in all cases fed by jets, although a sufficient flow of momentum is channelled to drive the outflows (Podio *et al.*, 2006). The accumulated moving mass in the bipolar outflows covers a wide range, from $10^{-2} M_\odot$ to $10^3 M_\odot$. Therefore, the mass set in motion can far exceed the mass accreted onto the protostar. In other words, we detect cold material which has been swept up or entrained by the jets.

The large-scale structure of protostellar jets has been intensively studied for a couple of decades. We will analyse a small selection of the well-known examples. The low extinction once the jet has exited the molecular core allows these jets again to be studied in the optical. However, it is the high mass flow which underlies the excited jets of molecular hydrogen that provides the major difference from the Class II jets (Davis *et al.*, 2010). The bases of these jets, called molecular hydrogen emission line regions, are revealed in near-infrared emission lines despite 5–50 mag of visible extinction.

These infrared jets are observed through their emission in lines of molecular hydrogen and atomic iron (Caratti o Garatti *et al.*, 2006). In particular, many high-resolution maps in the 2.12 μm 1-0 S(1) line (see Section 2.6) are now available. This line is produced when the molecules are vibrationally agitated, requiring shock waves of at least 9 km s^{-1} to be present. The shock waves heat the gas to ~ 2000 K in localised regions within the jet, generating multi-bow shock features, filamentary structures and more chaotic-looking distributions. The molecular jets rarely contain any smooth diffuse structure, but more often consist of a chain of arc-shaped aligned clumps separated on scales of between 1000 AU and 10000 AU.

The total emission in the molecular hydrogen outflow is high, with a value of $L_{H2} \sim 0.1 L_{bol}$, the bolometric luminosity of the source, for low-mass stars. There is also a remarkable correlation over five orders of magnitude in source luminosity of the form $L_{H2} \propto L_{bol}^{0.55}$. Moreover, the molecular component traced by H_2 lines is significantly enhanced in the youngest Class 0 protostars.

The jet density can be estimated from modelling the relative fluxes of several H_2 emission lines. Deriving the column densities in the upper energy levels of the radiative transitions, the collision frequencies can be determined, and hence the density of collision partners is found. This generally yields densities in the range 10^4–10^7 cm^{-3} on extended scales. However, these are densities of the shocked compressed gas, which occupies only a small part of the jet. Such high densities are, however, unavoidable for molecular jets from protostars, since the overall energetics require high densities. We thus obtain high mass ejection rates:

$$\dot{M}_{jet} = 1.3 \times 10^{-5} \left(\frac{r_j}{500\,\text{AU}} \right)^2 \left(\frac{n_j}{10^5\ \text{cm}^{-3}} \right) \left(\frac{v_j}{100\ \text{km s}^{-1}} \right) M_\odot\ \text{yr}^{-1}, \qquad (6.6)$$

where the hydrogen nucleon density is n_j and the molecular number density is $0.5 n_j$ (noting that molecular hydrogen is by far the most abundant molecule). Combined with the values for optical jets and microjets discussed above, this indicates an evolution in which the mass ejection rate falls as the accretion rate falls.

Jet speeds can be estimated through proper motions and spectroscopy. Several molecular jets display an almost linear increase in fluid speed with distance. Jet speeds inferred from proper motions of knots and radial speeds are generally quite low, between 40 and 100 km s^{-1}, in Class 0 jets. However, in the HH 111 and HH 121 jets, speeds above 400 km s^{-1} are found from both atomic and molecular components (Coppin *et al.*, 1998).

Atomic forbidden line emission in the form of infrared knots is also observed coincident with the major molecular knots within the jets from protostars. The main atomic tracers include [S II] and [Fe II] lines, which indicate that the jets are of low excitation typically produced behind low-speed shocks of speed 20–140 km s^{-1}. Spitzer Space Telescope observations of the L 1448-C jet reveal the existence of cool atomic gas at 2500 K, which may well be the major component of the jet (Dionatos *et al.*, 2009).

6.2.1 *HH 34 and HH 111*

The spectacular HH 34 jet is located in the L 1641 cloud in Orion at a distance of ~ 420 pc. As a Herbig–Haro object, the name HH 34 was given to the large bow shock to the south of what is now known to be the driving source HH-34 IRS (Fig. 6.6). Symmetrically placed to the north is a counterbow, clearly being driven in the opposite direction (Buehrke *et al.*, 1988). These are large, clumpy bow shocks, now called HH 34N and HH 34S, at a distance of 100" from the source. A number of other bow shocks and knots of emission on the same line have been detected, with the total extent of the HH 34 flow now recognised to be almost 3 pc. There is a large-scale S-shaped symmetry to the giant flow, and the jet displays a marked abrupt change in flow direction during a 65-year interval that ended 10 years ago. This could be interpreted as a disruption of the jet–disc system following the

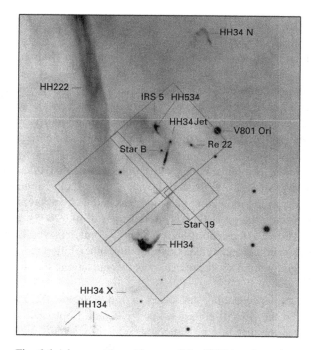

Fig. 6.6 A large-scale red image of the HH 34 region taken at the Cerro Tololo Inter-American Observatory 4 m telescope. The field is $\sim 4.5' \times 5'$. The dark lane at the top left of the image is emission from the peculiar HH 222. Smaller fields studied with the HST are outlined. Credit: Reipurth *et al.* (2002). Reproduced by permission of the AAS.

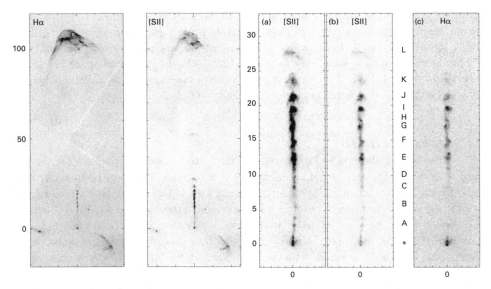

Fig. 6.7 The HH 34 jet. Left panels: the entire jet and bow shock as seen with the HST Wide Field Planetary Camera 2 through narrow-band filters which includes Hα 6563 Å(first panel) and [S II] 6730 Å(second panel). Right panels: close-up of just the jet in the two filters as indicated. Credit: Reipurth *et al.* (2002). Reproduced by permission of the AAS.

tidal effects associated with a recent periastron passage of a companion (Reipurth *et al.*, 2002).

The prominent highly collimated jet to the south is blue shifted, with a half-opening angle of just 0.4°. The first 30" consists of a chain of knots emitting in both optical (Fig. 6.7) and near-infrared lines. The HST images reveal that each knot has the morphology of a mini-bow shock with a bright [S II] core and a thin Hα filamentary arc where the shock extends into the ambient medium.

The counterjet is not detected in the optical but is seen at infrared wavelengths, as shown in the position–velocity diagrams of Fig. 6.8. Interestingly, an infrared reflection nebula is seen to the north rather than the south, indicating that more gas and dust are present to the north (but not sufficient to obscure the jet region). Garcia Lopez *et al.* (2008) detect the fainter red-shifted counterpart down to the central source. This jet contains several emission knots displaced symmetrically with respect to the corresponding blue-shifted gas, strongly suggesting an origin from an outbursting source which transmits pulsations simultaneously down both jets.

The radial velocities in the approaching jet cover a range from $-92 \, \text{km s}^{-1}$ to $-108 \, \text{km s}^{-1}$, as measured by the [Fe II] emission line by Garcia Lopez *et al.* (2008). A similar behaviour is found in the receding jet although, in addition, knots with significantly higher speeds are also found ($+140 \, \text{km s}^{-1}$), possibly due to variations in the jet axis projection angle. The H_2 radial velocities have a range from -89 to $\sim -110 \, \text{km s}^{-1}$ for the blue jet and an average value of $+115 \, \text{km s}^{-1}$ for the red jet, as shown in the position–velocity diagrams in Fig. 6.8.

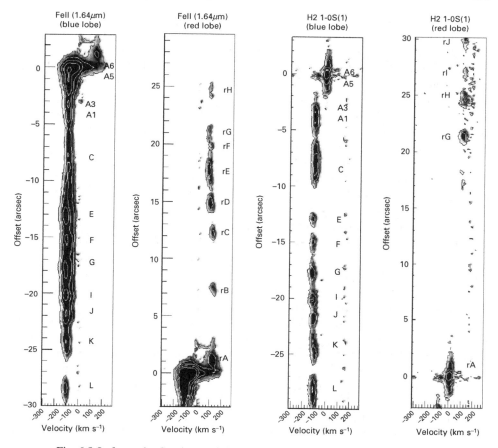

Fig. 6.8 Left panels: Continuum-subtracted position–velocity diagrams of the
[Fe II] 1.644 μm emission line for the blue and red jet of HH 34. A position angle of 15°
was chosen along the entire HH 34 jet. Right panels: the same for the H$_2$
1–0 S(1) 2.12 μm emission. Note that the contour levels are linear but not equivalent
between diagrams. Credit: Garcia Lopez *et al.* (2008).

The position–velocity diagrams and derived electron densities help constrain the magnetohydrodynamic launching mechanisms. While the kinematic characteristics of the line
emission at the jet base can be, at least qualitatively, reproduced by both X-winds and disc-
wind models, neither model explains the extent of the LVC and the velocity dependence of
the electron density (Garcia Lopez *et al.*, 2008). The LVC could instead represent denser
ambient gas that is entrained and partly accelerated by the high-velocity collimated jet.

Intrinsic line widths of the order of 35–40 km s^{-1} are observed all along the jet in the
[Fe II] emission line. This implies shock velocities of the order of 70–80 km s^{-1}, much
higher than the value of 30 km s^{-1} expected on the basis of the weak shocks corresponding
to the optical line ratios. Therefore, it seems that the line widening is determined not only
by the shock but also by, for example, a lateral expansion of the jet. In the inner jet region
of HH 34, components at high and low velocity (the so-called HVC and LVC) are detected

in both the atomic and the molecular gas. The [Fe II] LVC in HH 34 is detected to much larger distances from the source (further than 1000 AU) than in T Tauri jets.

In H_2, the LVC and HVC are spatially separated, with an abrupt transition from LVC to HVC at a distance of $\sim 2''$ from the star (Fig. 6.8), at variance with the [Fe II] result. This may indicate that the hydrogen molecules do not survive the highly excited, high-velocity inner jet environment. Reformation of molecules on dust grains within the jet could generate the HVC seen beyond 1000 AU. In this regard, careful measurements have concluded that dust in the jet is indeed still present (Podio *et al.*, 2006), as might be expected since the weak shocks should not completely destroy the dust.

Proper motions of the knots were first measured from the ground by Eisloeffel & Mundt (1992), while Reipurth *et al.* (2002) made very precise measurements with the HST. In the inner jet where a new knot was seen to emerge, a space velocity of at least 300 km s^{-1} is derived from the proper motions after correcting for the $60°$ angle of the flow to the line of sight. The jet rapidly slows down to a mean space velocity of about 220 km s^{-1} with a standard deviation of 20 km s^{-1} among the jet knots. The latter corresponds to internal motions within weak bow shock waves, consistent with the high ratio of [S II]/Hα in the jet. This is in accord with (1) steepening velocity oscillations which generate internal working surfaces and (2) the launch of high-velocity material which, perhaps through interaction with the dense ambient medium, rapidly slows down.

The high collimation is inconsistent with free expansion. If the jet has a larger internal pressure than its surroundings, it will expand freely at approximately the initial sound speed. A free expansion at an angle of just $0.4°$ with a bulk space velocity of 220 km s^{-1} corresponds to a radial expansion speed of just 1.5 km s^{-1}. However, the implied sound speed is extremely low, corresponding to a temperature of the jet gas of only 200 K (Reipurth *et al.*, 2002) whereas the optical spectrum would indicate 10^4 K. The problem can be resolved if the jet is enshrouded by a high-pressure confining sheath. The sheath itself need not be confined but can be supplied from the jet itself through the observed internal mini-bows, described as small internal working surfaces, which expel warm, shocked gas sideways at speeds of up to 20 km s^{-1}. In this manner, the observed jet regulates its own collimation through the pulsations.

The mass flux in both of the HH 34 jets is carried mainly by the high-velocity gas: lower limits on the mass flux of $3-8 \times 10^{-8} M_\odot$ yr^{-1} are estimated from the line luminosity (Podio *et al.*, 2006; Garcia Lopez *et al.*, 2008). The mass flux from HH34-IRS is only about 1% of the mass accretion rate of $4 \times 10^{-6} M_\odot$ yr^{-1} onto the growing star of mass 0.5 $M_\odot$ (Antoniucci *et al.*, 2008). The accretion luminosity of 13.3 $L_\odot$ provides about 80% of the bolometric luminosity, which may be typical of only the deepest embedded Class I sources. In more evolved objects, as they become exposed, accretion luminosities may fall to only a few per cent.

One of the finest known jets in star formation is HH 111, one of many discovered by Reipurth (1989). The outflow is located 420 pc away in the L 1617 cloud in Orion. The exciting source, IRAS 05491+0247, with a bolometric luminosity of $\sim 25 L_\odot$, is part of a hierarchical triple system still deeply embedded in the parental molecular cloud core. In fact, a second jet flow, HH 121, emerges from about the same position, suggesting that it is a binary.

The jet flows approximately east–west at an angle of only $11°$ to the plane of the sky, permitting accurate space motions to be derived from proper motions. The jet itself extends

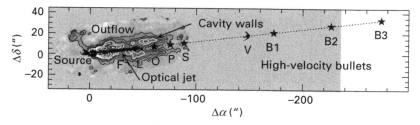

Fig. 6.9 The main features of the HH 111 blue jet and outflow superimposed on a map of the CO $J = 1 \rightarrow 0$ outflow emission (Lefloch *et al.*, 2007). Locations of the cavity, high-velocity CO bullets and optical jet (the shocked knots and the large bow shock V) are indicated. The CO is that integrated between +0.25 and +7.3 km s^{-1} obtained with the IRAM Plateau de Bure interferometer. Reproduced by permission of the AAS.

over several parsecs. Beyond the collimated jet, however, the flow continues and it is in fact only a minor part of a truly gigantic bipolar HH complex stretching over 7.7 pc (Reipurth *et al.* 1997a). The full extent of the blueshifted flow is displayed in Fig. 6.9.

The base of the jet is not visible at optical wavelengths. The main visible jet emerges from the cloud 15" from the source and is blueshifted. It resembles the HH 34 jet, consisting of a chain of numerous small knots with bow-like morphologies (Reipurth *et al.* 1997b) and emerging with higher velocities that rapidly drop to an approximately constant value. The jet terminates at a large bow shock (knots T-V) at a distance of 150". The redshifted jet is almost completely obscured, with only knot ZL visible in the optical while knots ZO and ZV (symmetric to knots O and T-V in the blue lobe) have been observed only in the near-infrared (Coppin *et al.*, 1998).

The HH 111 jet is a strong H_2 emitter and is surrounded by a powerful well-collimated CO outflow, whereas only very weak H_2 and CO outflows are associated with HH 34 (Chernin & Masson, 1995). In addition, three equidistant high-velocity CO bullets, with space velocities of about 240 km s^{-1}, are located further out (see Fig. 6.9, precisely along the flow axis (Cernicharo & Reipurth (1996)). These bullets probably represent working surfaces that no longer produce optical shock emission. The H_2 knots have now been observed in more detail by the Spitzer Space Telescope by Noriega-Crespo *et al.* (2011), which detects strong emission from the ground vibrational state as well as some emission from vibrationally excited states. The jet/counterjet system of knots displays a mirror symmetry that indicates a precessional motion of the ejection axis with a period of roughly 1800 years. This motion would be consistent with a binary of solar mass stars separated by $\sim 186\,\mathrm{AU}$.

The optical knots, as tracked on HST images of Hα and [S II], move ballistically (Fig. 6.10), i.e. directly away from the source with no evidence for turbulent motions (Hartigan *et al.*, 2001). Bow shocks have been observed to overtake each other while the wings of the bow shock L expand laterally. The proper motions translate accurately to space velocities in the range from 220 to 330 km s^{-1} with a typical uncertainty of just ± 5 km s^{-1}. The fastest knots are again associated with the base of the visible jet. Velocity differences between adjacent knots were found to be typically 40 km s^{-1}, consistent with maximum shock velocities of about 40 km s^{-1} as also inferred from the line fluxes (e.g. [O III] emission detected only in one location, at knot L).

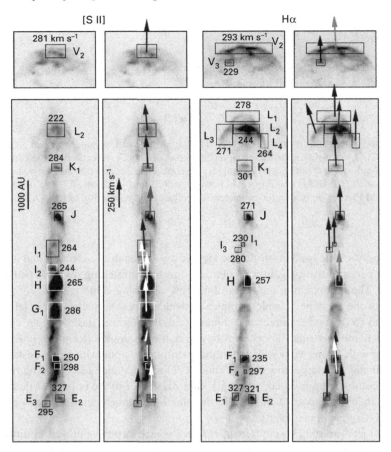

Fig. 6.10 The movement in the HH 111 jet calculated from [S II] and Hα HST images with a 4-year interval (for clarity, the arrows show the distance that would be travelled in 20 years). Derived speeds in the plane of the sky are in kilometres per second and correspond to the main feature in the box that defines the object. Motions perpendicular to the jet axis are small except for knot L3. Credit: Hartigan *et al.* (2001). Reproduced by permission of the AAS.

6.2.2 *Deeply embedded jets: HH 211 and HH 212*

Jets from Class 0 protostars reach the highest Mach numbers of all jets. They are extremely dense and powerful. The Class 0 source as well as the inner jet are obscured from view by a massive envelope which converts the accretion luminosity into far-infrared and submillimetre radiation, with Class 0 protostars being defined by a relatively high rate of submillimetre-to-bolometric luminosity (over 0.5%). Significant here is that the high rates of infall are matched by highly efficient outflows (Andre *et al.*, 1993). The jet mass flux is estimated to be $\sim 10^{-6} M_\odot$ yr^{-1}, which is probably 10–30% of the mass infall rate.

The extended jet structures are usually observable in the near-infrared. Hence, wide-field surveys for near-infrared jets are now a common means of detecting new Class 0 protostars (Davis *et al.*, 2009). Long before an outflow has accumulated a detectable amount of cool

CO gas, and before the protostar is detectable in the near-infrared or visible wavebands, the birth is already heralded by molecular hydrogen jets (Stanke *et al.*, 2000).

The infrared lines are produced in molecular shock waves in which the hydrogen molecule is vibrationally excited by collisions (Smith, 1994a). The total H_2 1-0 S(1) luminosity of the HH 212 flow is $6.9 \times 10^{-3} L_\odot$ in just the one line. The total H_2 emission in all lines is probably 20 times higher. Given the bolometric luminosity of the protostar of $\sim 14 L_\odot$, it is clear that a significant fraction of the accretion energy ends up in the jets.

The molecular hydrogen which produces the inner knots and bows has excitation temperatures of ~ 2400 K for the inner knots NK 1 and SK 1 and ~ 1700 K for the bows NB 1/2 and SB 1/2 observed further downwind. To be sufficiently excited to temperatures of at least 1000 K requires shock waves with speeds in excess of ~ 9 km s^{-1}. On the other hand, to remain sufficiently long in the excited state to radiate rather than dissociate requires a shock of speed lower than 23 km s^{-1}. However, the shielded environments of Class 0 jets may allow ambipolar diffusion to operate within the shock transitions (see Section 3.8). In the resulting magnetically mediated shocks, the shock energy is released mainly in the infrared lines and shock speeds can reach values of 50 km s^{-1} without dissociation of molecules (Draine *et al.*, 1983). However, these limits assume that the shock speed is at least an order of magnitude higher than the Alfvén speed. In jets and outflows driven by the magneto-centrifugal mechanism, the Alfvén speed is likely to remain very high even in extended jets. Magnetic cushioning may then extend these limits considerably (Smith *et al.*, 1991).

HH 212 is remarkable because of its high collimation and symmetrically placed pairs of bow shocks and shock knots on either side of the driving source (Zinnecker *et al.*, 1998). The near-perfect mirror symmetry displayed in Fig. 6.11 suggests that some outbursting event at the source itself is responsible for sending shock wave patterns at regular intervals down the jets. The symmetry is maintained even as the material streams through a dense gas core. The size of the outflow as measured to date is 240" or 0.54 pc.

The HH 212 jet source coincides with an infrared object IRAS 05413-0104, which is surrounded by a cold (12 K), compact core detected in ammonia emission (Fig. 1.3) from which it presumably accretes material. In the innermost region, water masers have been found in motion along the outflow (Claussen *et al.*, 1998). Proper motions of numerous maser spots within a projected distance of 40 AU are 30 ± 12 mas yr^{-1}, implying space velocities of 64 ± 27 km s^{-1}. An inclination of the outflow system to the plane of the sky of 4° is deduced from the relative magnitude of the proper motions and radial velocities of the masers.

On the next larger scale, the jet can be traced to within 500 AU through emission from the SiO molecule (Codella *et al.*, 2007). The high-velocity SiO gas is not tracing a wide-angle wind but is already confined to a flow inside a narrow cone of half-opening angle < 6°. SiO emission is not, however, associated with the inner pair of knots at a distance of 2500 AU, SK1 and NK1 (Lee *et al.*, 2007). Finally, at larger distances of 10" (5000 AU) and beyond, the spatial correspondence between the SiO and the H_2 is very good. The detection of [Fe II] emission in SK1 and NK1 and their 3–4 times broader line widths compared to other H_2 knots in HH 212 support the idea that these are stronger shocks. The series of inner knots are regularly spaced with inter-knot distances of 2000 AU. Emission is also detected in the inter-knot spaces: the knots appear to be connected by a thin stream of emission Both series of knots terminate in small bow shocks at a distance of about 13000 AU from the central

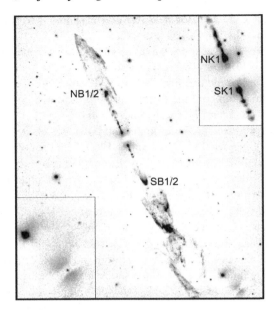

Fig. 6.11 The full HH 212 outflow in 2.12 μm H_2 emission as imaged by the ISAAC instrument on the Very Large Telescope in Chile between 2000 and 2002 shown in emission from H_2. The image size is $3.2' \times 3.9'$ or 3.7 pc $\times 4.5$ pc at a distance of 400 pc. Note that the continuum has not been subtracted so unrelated stars and galaxies may be apparent. Taken from McCaughrean *et al.* (2002).

position (NB1/2 and SB1/2). At high resolution, these bow shocks are seen to consist of two separate arcs nested together.

On close inspection there are some asymmetries in HH 212. Firstly, knot NK 1 displays a double-peaked H_2 line profile consistent with either a radiative bow shock or dual (forward and reverse) shocks. In contrast, the velocity distribution of the southern first knot SK 1 is single-peaked (Correia *et al.*, 2009), suggesting a lower jet velocity and possibly a different density variation in the jet pulses in the southern flow compared with the northern flow. Secondly, the two jet directions deviate by 2° from being antiparallel. Furthermore, there is a gradient in excitation transverse to the jet axis across the inner knots on the scale of 0.1" with a C-shaped inner symmetry that suggests a transverse source motion rather than jet precession (Smith *et al.*, 2007). This could be caused by a motion of the driving source relative to the cloud core or a motion of the cloud core relative to the external cloud medium (Fendt & Zinnecker, 1998). Alternatively, the jet direction could be influenced by the orbital motion of the jet source within a young binary system.

The HH 211 flow was discovered as a series of giant bow shocks and shell structure in H_2. This suggested the existence of a driving jet but, since the entire flow is deeply embedded, one was not discovered even in the near-infrared. Instead, as shown in Fig. 6.12, a high-speed jet of cold molecular gas was first revealed through submillimetre CO observations (Gueth & Guilloteau, 1999).

The jet has since been recovered in SiO, stemming ~ 100 AU from the apparent submillimetre source, as shown in panel (d) of Fig. 6.13 (Hirano *et al.*, 2006; Lee *et al.*, 2009).

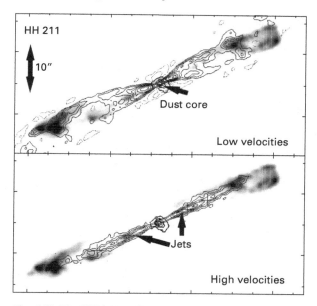

Fig. 6.12 The HH 211 outflow in CO $J = 2 \to 1$ rotational emission (contours) split into two radial velocity components. The top panel displays only speeds within 8 km s^{-1} of the maternal cloud while the lower panel displays the integrated emission for all other speeds. The H$_2$ near-infrared emission (greyscale) shows the warm shocked gas. Credit: Gueth & Guilloteau (1999).

The entire SiO jets, also seen in SO and CO transitions, consist of a chain of paired knots on either side of the source. The molecular abundances are enhanced, consistent with the release from dust grains and formation in internal shocks (Lee *et al.*, 2010).

At high resolution, the innermost pair of knots, BK1 and RK1, appear as two curved features with a C-shaped bending (reflection symmetry), connecting to the continuum emission. Knot BK1 extends to $\sim 0.3''$ (84 AU) from the source with a faint emission overlapping with the continuum emission and Knot RK1 extends to 140 AU to the west. These two knots are very narrow with a transverse width of under 40 AU (apparently containing sub-structure).

Proper motions of 170 ± 60 km s^{-1} for knots BK2, BK3, BK4, BK6, RK4, and RK7 were measured (Lee *et al.*, 2009). A possible velocity gradient is seen consistently across the inner pair of knots, 0.5 km s^{-1} across 10 AU, consistent with the sense of rotation of the envelope/disc. If this gradient is an upper limit of the true rotational gradient of the jet, then the jet carries away a very small amount of angular momentum corresponding to ~ 5 AU km s^{-1} and thus must be launched from the very inner edge of the disc near the co-rotation radius. Combining radial speeds with the proper motions yields an inclination angle of $5° \pm 2°$ and $6° \pm 2°$ to the plane of the sky for the eastern and western jet knots, respectively, and therefore a full jet velocity also of $\sim 170 \pm 60$ km s^{-1}.

The continuum source is now resolved into two sources, SMM1 and SMM2, with a separation of ~ 84 AU. The protostar could be the lowest-mass source known to have a collimated jet and a rotating flattened envelope-disc (Lee *et al.*, 2009). A small-scale 200 AU low-speed (2 km s^{-1}) collimated outflow is seen in HCO$^+$ around the jet axis extending

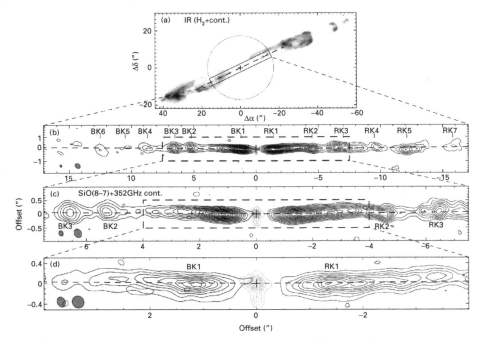

Fig. 6.13 The HH 211 H_2 outflow and SiO jet on three different scales. The cross marks the position of the protostar SMM1. Panel (a) displays the near-infrared image from Hirano *et al.* (2006). Panels (b) and (d) show the SiO emission as observed with the Submillimetre Array (SMA). The images are rotated by 266° clockwise. The redshifted emission on the right of SMM1 is integrated from 9.2 to 47.5 km s^{-1} and the blueshifted emission (east, to the left) is integrated from -21.2 to 9.2 km s^{-1}. Credit: Lee *et al.* (2009). Reproduced by permission of the AAS.

from the envelope-disc. It seems to rotate in the same direction as the envelope-disc and may carry away part of the angular momentum from it. Lee *et al.* (2010) suggest that the source could be a very low-mass protobinary with a total mass of $\sim 60\, M_{Jup}$ and a binary separation of ~ 5 AU.

Jets associated with more luminous and, hence, more massive young stars have proved elusive. Extremely well-collimated bipolar outflows which could be driven by jets have been associated with intermediate-mass stars, e.g. IRAS 20293+3952 (Beuther *et al.*, 2004). These examples may well correspond to scaled-up versions of low-mass star formation. It is likely that the majority of massive O stars also form by disc accretion. However, their outflows are broad and messy (e.g. DR 21, OMC 1, Cepheus A). It is plausible that the combination of a powerful stellar wind and radiative acceleration vacates the magnetic fields and material from the polar regions, thus leading to poorer collimation for their molecular outflows (Zinnecker & Yorke, 2007).

6.3   Termination: Herbig–Haro and molecular hydrogen objects

The location where a supersonic jet directly impinges on the ambient medium should be a site of strong shocks and intense radiation. The shock driven into the external medium may assume a roughly paraboloidal bow wave of extended proportions (Hartigan *et al.*, 1990). Rather than a single bow, a series of well-spaced bow-shaped structures is often detected. They are aligned along the jet axis on large scales, appearing to be driven either directly by the jets or as bow waves which lead the advance of free-floating supersonic bullets, e.g. in HH 240/HH 241 (O'Connell *et al.*, 2004). These bows can be found out to several parsecs, forming the giant parsec-scale flows (Reipurth *et al.* 1997a). The apices of the bows invariably face away from the protostar, giving the impression that they are being driven away from the forming star (Fig. 6.14). Convincing are the proper motions measured over time spans of a few years, which demonstrate that the bow structure as a whole moves away from the source often at a speed comparable with that of the jet itself.

The relevance to the theme of jets is most pressing in cases where the jet is seen to impact on the inner edge of the bow, as in HH 47 (Morse *et al.*, 1994). This occurs at the reverse shock, which brings the jet speed down to the speed of the interface with the ambient gas (Hartigan, 1989). The reverse shock front is predicted to take the form of a circular disc called a Mach disc (see Fig. 10.2). One such Mach disc was identified above in the HH 34 S bow shown in Fig. 6.7. Recalling that the HH 34 flow is inclined by 30° to the plane of the sky, an ellipsoidal structure might be discernible.

In the frame of reference moving with the bow, knots of emission move transverse to the bow structure away from the apex into the flanks (Eisloeffel & Mundt, 1994). In this co-moving frame, ambient gas enters from downstream and jet gas enters from upstream. The two gases shock, partly mix, and are then ejected sideways. This gas then encounters the ambient medium streaming by to produce the bow shape. The swept-up ambient material in the flanks is deflected at an oblique shock and flows back over the jet. The ejected mixture of ambient and jet material is diverted back to form a separating cocoon or sheath.

The working surfaces are much more complex than drawn in Fig. 10.2. The best-studied working surfaces of HH 34 and HH 47 (Hartigan *et al.*, 2005) are described as complex bow

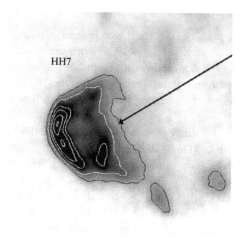

Fig. 6.14 The HH 7 bow shock observed in the molecular hydrogen emission line in the near-infrared at 2.12 μm (Smith *et al.*, 2003). The arrow denotes the direction of propagation from the driving star.

shock/Mach disc shock systems with numerous small clumps. The clumps have relative motions of up to 40 km s^{-1}. Clumps or instabilities continuously alter the Mach disc structure. With the bulk motion subtracted, the entire HH 34 Mach disc is then seen to gradually expand. The sideways expansion may reflect a recent event (roughly 200 years ago) when a high-velocity parcel of jet gas which reached the Mach disc was shocked and decelerated. It subsequently mixed with the ambient material that had passed through the bow shock and was shocked and accelerated. In addition, dense clumps may move all the way through the working surface to cause the bumpy morphology seen at the bow shock.

The leading (or forward) shock in HH 34 is split into a series of very thin, limb-brightened Hα-emitting filaments which envelope numerous knots (Reipurth *et al.*, 2002). One of these filaments developed four regularly spaced tiny knots in just a few years, possibly due to a fluid dynamic or thermal instability along the filament. Alternatively, overtaken clumps embedded within the ambient medium could provide ready-made obstacles. In some outflows including HH 34, slower bow shocks are found at larger distances from the driving source, thus displaying an apparent deceleration. This suggests that the bow shocks lose momentum progressively when they drift into the external medium.

The bow speed implied by the emission line spectrum is often significantly lower than the speed observed from the proper motion and deduced from the radial velocity (Davis *et al.*, 2010). If the shock is propagating rapidly it should be producing high-excitation atomic lines corresponding to a strong and fast shock. However, this is not always the case: despite high proper motions and radial velocities, many knots display a low-excitation spectrum corresponding to a low-speed shock. The only plausible resolution to this issue is that the bow is advancing through material which already has a substantial motion away from the source. In other words, many bows that we observe are not driven directly into the ambient cloud but are moving through the jet itself. This implies that outflows are larger and older than their present dynamical timescale and that, if we searched deep and wide, we would find evidence for even more distant HH objects. The discovery of these gigantic outflows has been made possible by developing detectors with the capability to image degree-scale fields (Davis *et al.*, 2007).

The chains of knots internal to jets are usually associated with internal working surfaces. In this case, the two weak shocks are called a reflected shock and a transmitted shock. They sandwich a growing layer of outflowing gas accumulated through both shocks. However, other mechanisms may generate similar patterns including damped Kelvin–Helmholtz instabilities which could form soft compressions of the central portion of the jet flow (Bacciotti *et al.*, 1995).

Molecular hydrogen objects also provide stringent tests for our understanding of molecular shock physics and molecular dynamics (Davis *et al.*, 1999). Some bows within jets such as HH 7 and HH 111 are detected in both molecular and atomic lines. It is then usually found that the molecular emission arises from the flanks of the bow, whereas the atomic emission originates from close to the apex (Fig. 6.15). This separation is expected since the molecules do not survive the high temperatures near the apex (Davis *et al.*, 1999). However, given bow speeds which typically exceed 60 km s^{-1}, the molecules should only survive in the far wings where the shock is sufficiently oblique that the component transverse to the shock surface is not dissociative, as illustrated in Fig. 6.15. One such example is in the S 233 H$_2$ bow shocks where a Mach disc is also identified in [Fe II] emission (Khanzadyan *et al.*, 2004).

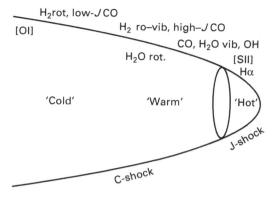

Fig. 6.15 The locations where radiative coolants make their maximum contribution are spread across the bow shock surface. The bow here is assumed to drive into a molecular cloud with a speed such that the molecules are dissociated within a cap. Hence, an atomic cap generates optical emission lines, while infrared fine-structure and rotational molecular emission are confined to the far flanks (Smith *et al.*, 2003).

In the case that the shock behaves hydrodynamically, molecules survive below the shock speed limit of ~ 23 km s^{-1}. In this so-called J-shock, there are two physical zones: a jump in temperature and density followed by a long cooling zone. In a J-shock, essentially all the heating of the molecules occurs within a narrow zone, a few collisional mean-free-paths wide (Section 3.7.3). Instead, we often measure H$_2$ emission from near the bow front (Smith *et al.*, 2003). This and the low excitation of the molecules suggest that the oncoming shock is being magnetically cushioned by ambipolar diffusion. With a low ionisation fraction, the momentum of the magnetic field spreads the front over a wide region within which radiative cooling almost keeps pace with the shock heating (see Section 3.8). As a result, thermal dissociation is suppressed. Furthermore, the shock fronts can be ~ 100 AU thick, which can be resolved. We may thus soon be able to observationally explore the flow pattern which ambipolar diffusion governs within these continuous C-shocks.

6.4   Bipolar outflows

As already noted, the jets may provide only a recent record of protostellar activity, whereas the slower-moving CO outflows hold the history. There are signs that the outflow was stronger and more collimated when the driving protostar was younger. The outflow is observed over an extended area, having a size range from under 0.1 pc to several parsecs (Bally & Lada, 1983), and exhibits bipolar structure with the redshifted emission formed in a different, possibly overlapping, region from the corresponding blueshifted emission. Analysis of the CO line emission indicates that typical flows have masses in the range $M \sim 0.3-100 M_\odot$. Therefore, the mass set in motion, especially in the youngest and most powerful outflows, far exceeds the mass accreted onto the protostar. We can conclude that while processes near the protostar drive a bipolar outflow, it does not provide the material. In other words, whereas the lobes of radio galaxies are formed out of jet waste, in bipolar outflows we observe material which has been swept up or entrained by the jet.

Outflow velocities are in the range ~ 10–$50\,\mathrm{km\,s^{-1}}$ and so lifetimes as distinctly recognizable dynamical entities of around $\sim 10^5$ years are computed, comparable to the age of a Class I protostar. It should be clear, however, that the kinematic age may not represent the true age. On the other hand, the dynamical time of even the longest jets is only a few thousand years.

6.5   Small-scale jets: radio and masers

Jets can still be detected through radio emission even in the vicinity of a dust-enshrouded young star (Anglada *et al.*, 1992). Observed at centimetre wavelengths on scales under 100 AU, the flux is often flat-spectrum or moderately rising with frequency (a positive radio spectral index). This testifies to the presence of aligned ionised gas jets producing radiation through the free-free mechanism (Section 2.4). The Very Large Array at centimetre wavelengths can resolve structure on the scale of 0.1" (Reipurth *et al.*, 2004) and high proper motions can be recovered (Marti *et al.*, 1998), although the radio continuum is often unresolved and more generally utilised to pinpoint the location and number of sources in a star formation region irrespective of the origin within a jet, wind, disc or binary companion (Reipurth *et al.*, 2004).

The process which ionises the jet can still be debated. It is attributed to stellar ultraviolet photons for high-luminosity objects, i.e. the latter stages of the formation of massive stars in which they have reached a final contraction stage and hydrogen fusion has begun or is about to begin. However, there are a large number of jets which are clearly not subject to a close source of substantial UV emission yet still produce weak (millijansky) radio emission (Anglada, 1996). For these, the ionisation appears to be shock-induced. In this case, a fraction of the jet momentum must be dissipated either through the steepening of pulsations into fast internal shocks or through interaction with the ambient gas envelope. Shock speeds exceeding $200\,\mathrm{km\,s^{-1}}$ will produce high levels of ionisation. In support, there is a rough linear relationship between observed centimetre radio power and the estimated rate of momentum outflow for a sample of young stars of low luminosity.

One of the most interesting thermal radio jets belongs to IRAS 16547-4247, a luminous young O8 zero-age main-sequence star with a bolometric luminosity of $6.2 \times 10^4 L_\odot$ at a distance of 2.9 kpc (Garay *et al.*, 2003). There is a string of infrared H_2 knots on the parsec scale and a highly energetic collimated bipolar outflow with lobes extending to 0.6 pc in opposite directions, centred on the infrared source. This is evidence that jets are not excluded from high-mass young stars. Rodríguez *et al.* (2008) demonstrate that the jet phenomenon associated with such massive YSOs could be a scaled-up version of the low-mass case.

A centrally located radio source on the arcsecond scale corresponds to the thermal jet, with an opening angle of 25° as shown in Fig. 6.16. However, there are two outer radio lobes about 8" away described as radio Herbig–Haro objects (Brooks *et al.*, 2007). Each lobe contains both thermal and non-thermal synchrotron emission components. This is probably consistent with a fast shock wave, perhaps $500\,\mathrm{km\,s^{-1}}$, in which a fraction of the electrons are accelerated to relativistic speeds via the Fermi process as they are scattered many times across the front. Propagation through a magnetised medium then generates synchrotron emission, while most of the electrons remain thermal to produce the free-free component.

H_2O masers at 22 GHz are frequently detected in low-mass young stars and are known to be a good tracer of jet activity very close to embedded protostars, while other masers

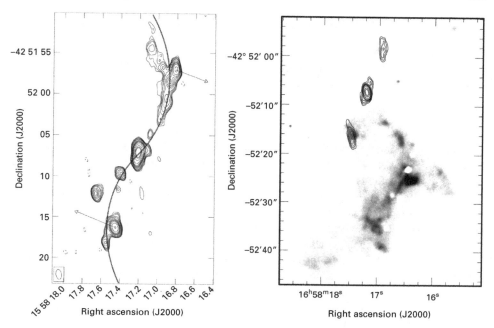

Fig. 6.16 Left: A radio VLA contour image at 8.46 GHz toward IRAS 16547-4247 for epoch 2003.74. The solid line depicts a spiral model discussed by Rodríguez *et al.* (2008). The arrows indicate the proper motions of components N-1 and S-1 for a (supposed) period of 300 years. Right: the VLT H_2 2.12 μm emission overlaid with 25 GHz emission contours from ATCA. Credit: Rodríguez *et al.* (2008) and Brooks *et al.* (2007). Reproduced by permission of the AAS.

are associated with protostellar discs. The H_2O masers are excited in gas with temperatures of ~ 400 K and number densities of 10^8–10^{10} cm^{-3} (Elitzur, 1992). The masers are highly time variable on timescales of a day to a month with luminosities that correlate well with the luminosities of the radio continuum emission, also suggesting that masers trace the jet activity. They both appear to be excited in the shocked layer between the ambient protostellar core and jet, although the origin is not well understood.

Masers provide a means to probe the launch site of obscured jets. An extremely well-collimated maser jet has been found close to the launch site during an exploration of the nearby region NGC 1333-IRAS 4. It was detected in the water line with the Very Long Baseline Array by Desmurs *et al.* (2009). For IRAS 4B the red- and blueshifted lobes very close to the protostar, within 35 AU, were found to be moving away with projected velocities of ~ 10–50 km s^{-1}. Desmurs *et al.* (2009) estimate a jet width of just 2 AU at a distance of 12 AU from the driving protostar.

6.6   Summary

The reason why YSO jets are present has not been resolved. As we have discussed, the presence of rotation in jets has provided controversial results. The extraction of angular momentum remains to be confirmed. This is not surprising, however, since jets are high-speed and probably do originate from the inner edges of the accretion disc where almost all of the angular momentum has already been lost through other mechanisms such as planet formation and winds.

According to most reports, jets are not important as a feedback of energy into the molecular cloud. Turbulence and mass dispersal may be effective from the core on a scale of under 1000 AU, but how a collimated jet does more than puncture a small cavity, and race away ballistically through the cloud, remains to be shown. Jet precession through a wide angle may provide a solution.

Jets are hypersonic. However, optical jets possess high internal sound speeds of order 10 km s^{-1}, which should correspond to their transverse speed of expansion unless they are confined. It is not clear that jets expand with the required opening angle, often of several degrees. Magnetic fields in the ambient medium may be present to achieve the pressure on the jet itself or the jet may expel high-temperature gas through internal bow shocks and so be self-collimating by forming its own sheath. High magnetic fields within the jet are also expected within all magnetocentrifugal models and the dynamical effects of the field should be critical to the knots and bows which usually dominate the jet appearance.

Ideally, we will be able to resolve the launch site; or, at least, resolve where the jet collimates close to the star. Discs are typically 100 AU in size but the collimation occurs on smaller scales. We will need to resolve the 10 AU scale, requiring well under 0.1" spatial resolution for even the closest sources. This is difficult to achieve from the ground, especially in the optical, where the brightest lines emit. Many jets are unsuitable due to obscuration from the disc and core and also because associated phenomena such as bow shocks and slow winds complicate measurements.

7

Jets associated with evolved stars

This chapter presents a mixed bag of objects and phenomena from planetary nebula to gamma-ray bursts. The data are fragmentary with results often representing the behaviour of selected attractive objects which may be contradicted by further observations. Nevertheless, the importance of stellar jets to the overall progress in the topic is now immense. This promotion has been earned by the short timescales involved and the known properties of the driving sources. This provides opportunities to track changes in jets, which are often transient, and to relate these changes to the launching accretion disc and star.

The evolutionary timescales associated with accretion discs and jets should scale with the luminosity and mass of the central object. Therefore, the changes that would take millions of years in quasars should occur over just hours and days in galactic sources. Much depends on the escape speed from near the surface of the particular star. Symbiotics and supersoft sources are accreting white dwarfs. Microquasars are radio-emitting X-ray binaries with a radio morphology like quasars and high X-ray luminosity. The primary can be a neutron star or black hole. Gamma-ray bursts are associated with collapsars or hypernovae, which generate black holes and ultra-relativistic jets.

For each set of objects, we define (1) the driving set-up, i.e. the stellar system; (2) the major discoveries; (3) the source activity behind or accompanying the jet launching, e.g. outbursts; and (4) the jet phenomena. The review strategy assumed here is one of speed: we begin with the slower jets and finish with superluminal motions.

7.1 Planetary nebulae

Stars which have evolved onto the asymptotic giant branch of the Hertzsprung–Russell diagram are immersed in roughly spherical gas-dust envelopes. The envelopes were previously ejected in the form of slow, dense winds at the end of their evolution as low-mass and intermediate-mass stars. The winds have mass loss rates of up to $10^{-4} M_\odot$ yr^{-1}. Associated with *post*-asymptotic giant branch (post-AGB) stars are proto- or pre-planetary nebulae (PPNs). The PPNs will develop into planetary nebulae (PNs) in less than 1000 years as the post-AGB stars evolve into hot white dwarfs. With this brief duration, they are rare objects to catch.

In the short transition period between the AGB and PN phases, the PPNs turn on very fast winds with speeds of a few 100 km s^{-1} to 2000 km s^{-1}. The winds interact violently with the surrounding envelopes and drastically modify their spatial and kinematic structures. Specifically, a significant fraction of PPNs and young PNs take up the form of highly

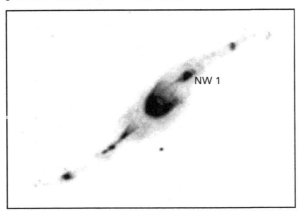

Fig. 7.1 Henize 3-1475. The protoplanetary nebula He 3-1475. imaged with the Hubble Space Telescope WFPC2 through the 6580 Å [N II] narrow-band filter. The image size is 15.9" × 11.4"; at an assumed distance of 5 kpc, the jets extend ∼0.1 pc. The Knot NW1 is the location of an X-ray source. North is up, and east is to the left. The bases of the jets are widely separated from the central star, a common feature of PPN jets. Credit: Borkowski *et al.* (1997). Reproduced by permission of the AAS.

collimated bipolar or multipolar lobes. The shaping mechanism is still generally unclear and is closely related to the mass-loss processes. The final extended planetary nebulae show a dazzling variety of morphologies.

One of the finest pairs of jets in a proto-planetary nebula belongs to Henize 3-1475, displayed in Fig. 7.1. The nebula itself is highly collimated and bipolar. It contains S-shaped jets stemming from the vicinity of a massive post-AGB star (Borkowski *et al.*, 1997). The jets are a string of optical knots extending over 17" along its main axis with point-symmetry. The spectra of the knots indicate shock excitation with bow shocks of speed 150–200 km s^{-1}(Riera *et al.*, 2006). Radial velocities are much larger, suggesting that there is a continuous stream within which the shock waves propagate. All these results are consistent with episodic or pulsed jets whose axis has precessed with a period of about 1500 years through an angle of under 10°.

X-ray emission associated with a Henize 3-1475 jet was discovered from the Chandra X-ray Observatory by Sahai *et al.* (2003). The compact X-ray feature is coincident with the brightest optical knot, NW1, which is displaced by 2.7" from the star. The peculiar apparent narrowing of the flow at this knot is not seen in other jet-producing environments. It could be a potential example of reconfinement or a 'conical converging flow', in which a jet flow is focused by the walls of an axisymmetric cavity to converge towards a point where shock waves deflect the material towards the jet axis.

The jets are likely to be the result of (re-)accretion of circumstellar material within a binary system. This scenario explains very young PPNs, in which the central star is too cool to exhibit a line-driven wind with sufficient momentum. PPNs raise the issue of the difference between explosive jet launching, such as those discussed in the context of gamma-ray bursts and X-ray transients, and true streaming in which the flow continues for a time exceeding the dynamical flow timescale. Dennis *et al.* (2008) argue that the clump model is favoured over continuous streams because it can produce higher collimation and account better and more naturally for ring-like structures such as observed in the PPN CRL 618. HST images of CRL 618 have revealed the presence of narrow lobes at different orientations. Several bow-like structures are seen within the body of the lobes and two bow-shaped structures observed at the tips of the lobes. It is difficult to conceive how even a precessing jet could produce simultaneous ejections in three directions (Trammell & Goodrich, 2002).

Other PPNs show strong evidence for collimated but precessing outflows, although jets have not been observed. The Water Fountain Nebula, IRAS 16342-3814, displays a bipolar reflection nebula with a dark equatorial waist on Hubble Space Telescope images (Sahai *et al.*, 1999). This is interpreted as an optically thick dusty torus which completely obscures the central star. In the near-infrared, a striking corkscrew structure is identified which can be best explained as a precessing jet with a diameter of 100 AU and a precession period of 50 years (Sahai *et al.*, 2005). Such corkscrews are found in a few PPNs and are termed, or rather interpreted in terms of, bipolar episodic rotating jets (BRETs). The PPN inherits its name from remarkable H_2O maser activity. The emission exhibits widely separated doublets with a velocity spread of $259 \, km \, s^{-1}$. The masers are therefore interpreted as streams or clumps of molecular gas forced out along the polar axes (Likkel & Morris, 1988).

A number of PPNs display collimated outflows that originate near the central star and impact the bipolar lobes. These could be attributed to jets in the broadest definition. Examples of collimated and curved flows of clumps and shock waves include the near-infrared M1-16 flow (Aspin *et al.*, 1993) and the optical M1-92 flow (Trammell & Goodrich, 1996). Similarly, the so-called jet within the northern lobe of the Rotten Egg Nebula, OH 231.8+4.2, takes a rather flocculent appearance with a sharp edge on one side (Meakin *et al.*, 2003). The original Egg Nebula, CRL 2688, is one of a class of young planetary nebula that possess multiple collimated outflows which are purported to be driven by multipolar jets (Cox *et al.*, 2000). The planetary nebula K1-2 contains a collimated string of low-ionisation knots embedded in an elliptical shell (Corradi *et al.*, 1999). The knots expand with a velocity similar to that of the elliptical nebula ($25 \, km \, s^{-1}$) with an extended tail protruding out of the main nebula. It exhibits a linear acceleration up to $\sim 45 \, km \, s^{-1}$ with the ejections from a close binary system that probably underwent a common-envelope phase during the AGB stage. Although commonly discussed as being driven by spectacular jets, these are all better described as collimated flows.

Mature planetary nebulae contain widespread small-scale structures (e.g. the Helix Nebula, NGC 7293) as well as specific features in the form of patches, bullets, knots and filaments which generate low-ionisation shock-excited optical and infrared emission lines. In some objects, collimation is apparent in the form of bipolar ansae or axial structures, which hint at the presence of jets. Pairs or sets of knots lying along or near the apparent symmetry axes are not unusual. In addition, thin finger-shaped lobes are sometimes observed e.g. OH 231.8+4.2 (Bujarrabal *et al.*, 2002). Henize 401 is a similar bipolar planetary nebula with highly collimated lobes suspected to result from the momentum-driven shock wave interaction of a high-velocity bipolar jet with the progenitor circumstellar envelope (García-Lario *et al.*, 1999).

In other objects, bright features near the minor axes indicate the presence of discs or tori. Either by collimating a fast stellar wind or by driving a jet via accretion in the central system, dusty tori or stable discs may be crucial ingredients for the shaping of planetary nebulae. Fully articulated jets, however, are very rare. Thus, the creation of continuous jets as in the case of YSOs and radio galaxies does not seem to be the norm in PNs and PPNs.

The outflow speeds of most highly collimated planetary nebula are often hundreds of km s^{-1}, more than a factor of 10 larger than the escape velocity from the surface of an AGB or post-AGB star. For example, extremely high speeds in remote knots along the symmetry axis are found in the Engraved Hourglass Planetary Nebula MyCn18 (O'Connor et al., 2000). These knots possess a range of outflowing speeds that are proportional to their distance from the central star, reaching up to 630 km s^{-1}. There is a degree of point and velocity symmetry which indicates that some pairs of knots were ejected in opposing directions at the same speed.

The high observed speeds can readily be obtained in close binary systems with orbital periods in the range of a few days to few times 10 years (Soker, 1998). Furthermore, binary systems can induce an accretion disc precession and, hence, a jet precession. Precessing jet models have been applied to PNs and are also relevant to PPNs such as Henize 3-1475. There are many point-symmetric planetary nebulae including the remarkable double-S shape of IC 4634 (Guerrero et al., 2008), which contains an arc that could be the relic of ancient precessing collimated ejections, as well as inner S-shaped features. Hence, a binary companion is increasingly gaining support as a dominant shaping mechanism for many planetary nebula.

7.2   Symbiotic systems

A symbiotic star is a variable binary system in which one star has expanded its outer envelope and is rapidly shedding mass and another hot star is ionising this gas. Most symbiotics consist of a white dwarf which is accreting material from the wind of a red giant. The symbiotic system originates from a system with mass of the order of $5\,M_\odot$. The more massive star evolves faster through the main sequence and then, as a red giant, sheds most of its mass through a stellar wind. The expelled material can occasionally be detected in the extended environment. This star becomes the hot white dwarf. In the meantime, the less massive star has retained its mass and eventually itself enters the red giant phase, experiencing large mass loss.

More than 200 symbiotic stars are known and at least ten systems display jet phenomena, the best-studied of these systems perhaps being CH Cyg, R Aqr, RS Oph, Z And and MWC 560.

It is not clear that the collimated flows are true jets. The plasma may emerge geyser-like in twisted streams invariably associated with optical eruptions. The force of the explosion may rather eject bullets which are channelled outwards and confined by strong magnetic fields. The jets are imaged in the radio and inferred from optical narrow emission lines displaced in radial velocity. The nature of the outbursts recorded for the majority of classical symbiotics remains debated, with disc instabilities, nuclear flashes in a shell or photosphere expansion invoked (Brocksopp et al., 2004). On the other hand, the slow novae and recurrent nova symbiotics are thought to be thermonuclear runaways, although it not always clear for the latter. The outburst duration and recurrence vary widely, between months and decades.

The CH Cygni, system ejected a double-sided jet with multiple components during a strong radio outburst in 1984/1985 (Taylor et al., 1986). The onset of the radio outburst coincided with a remarkable decline in the visual brightness of the star. Changes in the fast optical flickering in 1997, after a jet was produced, have also suggested a direct connection between the accretion disc and the production of jets (Sokoloski & Kenyon, 2003). The

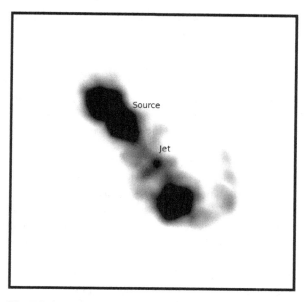

Fig. 7.2 CH Cygni. A Hubble Space Telescope image of the jet-like structure in the symbiotic CH Cygni. The F502N filter contains the 5007 Å line of [O III]. The image size is 4.5" × 4.5". Image provided by M. Karovska, see Karovska *et al.* (2010). The counterjet to the north-east is barely visible.

radio jet has been resolved with a width of 1" and length 3" at a distance of just 250 pc. It exhibits a roughly flat gigahertz spectrum consistent with optically thin free-free emission (Karovska *et al.*, 2010). Karovska *et al.* (2010) were able to estimate the ionised gas density at 1.5×10^4 cm^{-3} on taking a temperature of 10^4 K. They also found the prominent optical jet shown in Fig. 7.2.

CH Cygni contains X-ray emission associated with the southern jet (Galloway & Sokoloski, 2004) with Chandra detecting a component at 1.5" (~ 375 AU) from the central source. Along with R Aquarii, the thermal X-rays appear to be produced where jet material is shock-heated on collision with the circumstellar gas. An optical bipolar outflow on the scale of 5000 AU could well be the remnant of the interaction of the bullets with this relatively dense ambient medium (Corradi *et al.*, 2001). On the other hand, non-thermal (synchrotron) radio emission is associated with the extended jet region, possibly partly caused by magnetic field enhancements in compacted knots of the order of 3 mG (Crocker *et al.*, 2001).

The CH Cygni jet position angle on the sky changes with time, consistent with jet precession. A precession period of 6520 days within a precession cone opening angle of 35°and a jet velocity of 1500 km s^{-1} were consistent with the data (Crocker *et al.*, 2002). Subsequently, the shock wave front has slowed to a speed under 100 km s^{-1} after having moved from 300 AU to 1400 AU to the south. In addition, a new jet in the NE–SW direction has been detected by Karovska *et al.* (2010) in the X-ray, optical and radio. This new jet may be again related to a source decline, with a strong dimming of CH Cyg in the second half of 2006. These are clumpy jets with components extending out to ~ 750 AU. The structure suggests some combination of precession, pulsation and outbursts. The large precession angle may be related to the misalignment between the jets and the orbital spin axis of the binary system as resolved by Mikołajewska *et al.* (2010).

The symbiotic prototype Z And revealed bipolar jets that appeared and disappeared during its 2006 outburst (Skopal *et al.*, 2009). They were launched asymmetrically with a red:blue velocity ratio of 1.2–1.3. They became symmetric from about mid-August onward at 1200 km s^{-1}, and the velocity then fell to 1100 km s^{-1} at their disappearance. The spectral properties of these satellite emission lines indicated the ejection of bipolar jets collimated within an average opening angle of 6°.

If the jets were expelled at the escape velocity, then the mass of the driving white dwarf is 0.64 $M_\odot$. The average outflow rate of mass through the jet is estimated to be $\dot{M}_{jet} \sim$ $2 \times 10^{-6} (R_{jet}/1 \text{ AU})^{1/2} M_\odot \text{ yr}^{-1}$, during their August-September maximum. During their lifetime, the jets carried a total mass of $\sim 7.4 \times 10^{-7} M_\odot$. Evolution in the rapid photometric variability and asymmetric ejection of jets around the optical maximum was interpreted as due to the disruption of the inner parts of the accretion disc caused by radiation-induced warping.

The jet axis in MWC 560 is close to parallel to the line of sight (Tomov *et al.*, 1990). This special orientation provides the opportunity to observe the outflowing gas as line absorption with radial velocities as large as 6,000 km s^{-1}, in the source spectrum. Therefore, MWC 560 can be used to probe the short-term evolution and the propagation of the gas outflow in jets from white dwarfs. Detailed modelling (Stute *et al.* 2005a) summarised below provides some good quantification of the essential parameters in a symbiotic jet. A later confirmed prediction of soft X-ray emission but at a much lower flux level indicates a rapidly variable jet flow (Stute & Sahai, 2009).

R Aquarii is one of the nearest symbiotic stars with a late-type giant of the Mira type and a white dwarf companion. It is a well-known jet source (Herbig, 1980; Nichols & Slavin, 2009) first noticed as a spike in 1977 by Wallerstein & Greenstein (1980). At a distance of about 200 pc, the jet-like features can be imaged over a distance of at least 2500 AU on both sides of the source.

The outer thermal X-ray lobe and radio jets have been followed by Chandra and VLA observations separated by 4 years (Kellogg *et al.*, 2007), shown in Fig. 7.3. An advancing proper motion of only 580 km s^{-1} was uncovered in the X-ray emission from the north-east outer X-ray lobe (away from the central binary). The south-west outer X-ray lobe almost disappeared between the two dates. This could be due to adiabatic expansion and cooling while the north-west jet continues to interact with much denser surroundings. The X-ray emission originates from the outer jets, while optical and radio dominate the inner jets. The south-west jet is not antiparallel to the north-east jet but is around 45° out of alignment, suggesting precession, rotation or deflection.

The presence of jets is often taken to imply that accretion is ongoing and achieved through a disc. The presence of a disc is also inferred in R Aqr from strong, short-term (hour) flickering of the hot component. In addition, the central object is known to undergo irregular outbursts, with the first recorded outburst in the year 1073.

Radio observations found continuum emission approximately co-spatial with the optical jet (Sopka *et al.*, 1982). The radio jet emission was later resolved into a series of knots which trail the optical emission rather than being co-spatial as if the jet were precessing (Hollis & Michalitsianos, 1993). The displacement is consistent with the optical emission preceding

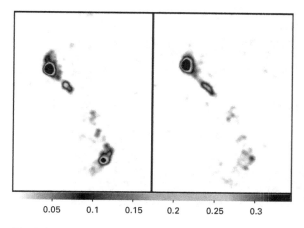

0.05   0.1   0.15   0.2   0.25   0.3

Fig. 7.3 R Aquarii. Chandra X-ray images of R Aqr from the epochs 2000.7 (left) and 2004.0 (right) with X-ray counts in the energy range 0.2–3.5 keV. Note that the SW jet faded almost beyond detection in the 3.3 years between these observations. The NE jet X-ray emission has advanced from the central binary with a projected velocity of about 580 km s^{-1} in the sky plane (Kellogg *et al.*, 2007). Reproduced by permission of the AAS.

the compressed recombination zone which generates the radio free-free emission. The jets are embedded in an arcminute-scale outer nebula containing an elliptical inner nebula consistent with formation and inflation through multiple ejection events roughly 180 and 600 years ago. The Hubble Space Telescope also showed numerous linear filamentary features of ionised gas emanating from the inner core. The jet itself radiates after excitement through photoionisation near the source, with shock waves around or in clumps further out. The line profiles are consistent with advancing bow shock waves (driven by quite dense jets or bullets) rather than shocked cloudlets (obstacle has zero velocity relative to the source). Synchrotron radiation from the new jet in contemporaneous VLA radio spectra implies that at physical conditions in the early stages of jet development are different from those in the more extended outer thermal jets known to exist for decades in this system (Nichols *et al.*, 2007).

The final example presented here is the symbiotic and recurrent nova RS Ophiuchi. It experiences nova eruptions approximately each 20 years, with the latest eruption having occurred in February 2006. Radio synchrotron jets in the east–west direction have been associated with both this and the previous 1985 eruption.

The radio image of Sokoloski *et al.* (2008) displayed here in Fig. 7.4 shows that the eruption in 2006 produced thermal free-free jets with half-opening angles of just a few degrees that supply extended lobes of relativistic, synchrotron-emitting particles (contours). Assuming a uniform jet velocity, the jets persisted for more than a month after the start of the explosion. It appears that the explosion did not completely destroy the accretion disc or that an accretion disc is not needed to collimate the jets. King & Pringle (2009) also note that the disc should be expelled from the white dwarf vicinity given that the energy yield greatly exceeds the disc binding energy. Instead, it was suggested that the outbursts are created in the disc by thermal-viscous instabilities with the disc irradiated by the central accreting white dwarf.

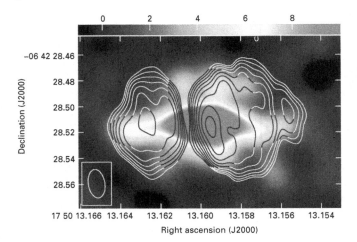

Fig. 7.4 The RS Oph jets and lobes. The image shows the high-frequency 43 GHz radio emission as imaged by the VLA, which preferentially traces thermal plasma. The contour lines show the lower-frequency 1.7 GHz emission as imaged by the VLBA, which preferentially traces synchrotron-emitting gas. Image taken from Sokoloski *et al.* (2008). Reproduced by permission of the AAS.

The origin of symbiotic jets remains controversial (Sokoloski *et al.*, 2008). Proposed mechanisms include (1) intrinsically asymmetric explosions, (2) discrete ejecta that move into an inhomogeneous environment, and (3) continuous driving and collimation similar to those from black holes and protostars. The latter, from a precessing accreting disc, would appear most plausible.

For jet propagation, hydrodynamical simulations with parameters representative of the symbiotic systems have proved successful (Stute *et al.* 2005a). Simulations of a pulsed, initially underdense jet in a high-density ambient medium were performed in a cylindrical coordinate system to model MWC 560 specifically. The jet was produced within a thin nozzle with a radius of 1 AU and a binary separation of 4 AU. The red giant generates a stellar wind with constant velocity of $10 \, \text{km s}^{-1}$ and a mass-loss rate of $10^{-6} \, M_\odot \, \text{yr}^{-1}$, modelled as originating from a time-averaged ring. Pulses were modelled as impulsive $200 \, \text{km s}^{-1}$ increases on each 7th day, gradually decreasing over several days superimposed on a steady jet of $1000 \, \text{km s}^{-1}$ and density $5 \times 10^6 \, \text{cm}^{-3}$. These parameters lead to a mass-loss rate of $10^{-8} \, M_\odot \, \text{yr}^{-1}$ in the steady jet, a density contrast between the steady jet and the ambient medium of 5×10^{-3}, and a Mach number of 60 at the jet nozzle.

Symbiotic jets remain not so well investigated as other objects. Due to the fact that their parameters are in a different regime, their study may yet provide unique insight. The jets are probably underdense with a density contrast of 0.001–0.01 (compared with 1–10 in YSOs, 0.1 or lower in AGN). The jet speeds are typically 1000–$5000 \, \text{km s}^{-1}$, a factor of 10 higher than in YSOs but up to 100 times lower than in AGN (Stute *et al.* 2005a). The absolute densities of 10^6–$10^8 \, \text{cm}^{-3}$ are similar to YSO jets at the equivalent jet length but considerably smaller than the AGN jet values of $10^{-2} \, \text{cm}^{-3}$. Therefore, symbiotic jets may be underdense but with strong radiative cooling. We remark that similar precessing jet-like outflows have been found in pre-planetary and planetary nebulae, a fraction of which probably evolve from symbiotic systems (e.g. Sahai *et al.*, 2005).

7.3 Supersoft X-ray sources

A supersoft X-ray source (SSS), or compact binary supersoft X-ray source, emits strongly in the low-energy soft X-ray range (roughly 0.1–10 keV, with hard X-rays in the 10–100 keV range). Luminous galactic SSSs have characteristic luminosities of 10^{36}–10^{38} erg s^{-1} and effective temperatures of 2–6×10^5 K (Becker *et al.*, 1998). The soft X-rays are produced through steady nuclear fusion on or just below the surface of a (massive) white dwarf (van den Heuvel *et al.*, 1992). To achieve steady fusion requires a high accretion rate. This material has been pulled from a binary companion, via an accretion disc before being deposited onto the white dwarf. It is important that the flow must be rapid ($10^{-7} M_{\odot}$ yr^{-1}) to sustain the continuous stable fusion. As the dwarf gains weight, it may eventually become unstable, igniting carbon upon reaching the Chandrasekhar mass, at which point it explodes as a Type Ia supernova, without leaving behind a core.

Supersoft sources are mainly detected in nearby external galaxies at relatively high galactic latitudes such as the Large Magellanic Cloud (LMC) ($-33°$) and M31 ($-21°$). In the plane of our own galaxy, a hydrogen column density of only 10^{21} cm^{-2} suffices to absorb the very soft X-radiation. In the galactic plane this corresponds to distances of less than a kiloparsec. Some of these sources have shown indications of jets through the appearance of 'satellite' emission lines in their spectra. These features are transient with typical timescales of months. For the galactic SSS counterparts, termed V Sagittae (V Sge) stars, such features have been reported for WX Cen and V617 Sgr (Steiner *et al.*, 2007) with a large difference in radial velocity of Balmer line satellites at high and low photometric states. Given an inclination angle of 72° for the V617 Sgr system, the intrinsic speed of the jets is inferred to be 2500 km s^{-1} when they were as shown in Fig. 7.5. However, the lines vary in width and separation, leading to a debate on their interpretation as bona fide jets.

Low-velocity, bipolar jets were also discovered in the galactic source RX J0019.8+2156 (QR Andromedae) by Becker *et al.* (1998). These transient jets are inferred from redshifted-blueshifted pairs of emission lines of H and He II with a radial component of outflow velocity of 815 km s^{-1}. When present, the jet lines seen in Hα also exhibit a modulation of 71 km s^{-1} due to the orbital velocity of the white dwarf, which also indicates that the jets are oriented nearly perpendicular to the orbital plane. Note that the profiles of the jet lines possess an intrinsic FWHM of 400 km s^{-1}, suggesting very low collimation or other broadening mechanism.

The three highest-luminosity sources show evidence of jet outflows, with velocities of $\sim$1000–4000 km s^{-1} (Cowley *et al.*, 1998) consistent with the escape velocity from a white dwarf. A high inclination angle is responsible for the low radial velocity of QR And and V617 Sgr (Steiner *et al.*, 2007). In CAL 83, the shape of the He II 4686 Å profile shows evidence for jet precession with a period of $\sim$69 days. Southwell *et al.* (1997) interpreted this as due to a warping instability of the accretion disc, which is strongly irradiated by the central object.

Transient jets were also observed in the supersoft X-ray source RX J0925-4758 (MR Vel) (Motch, 1998). The jet was detected in the Hα line with a peak jet velocity of 5200 km s^{-1} during a single night in June 1997 and had disappeared one day later. Simple modelling of the jet profile would suggest a half opening angle in the range 17°–41° although once gain the outflow may be narrower if part of the observed spread in velocity is intrinsic to the jet. The system remains controversial with a high escape speed, high inclination angle and the unusually long orbital period of near 3.8 days (Schmidtke *et al.*, 2000).

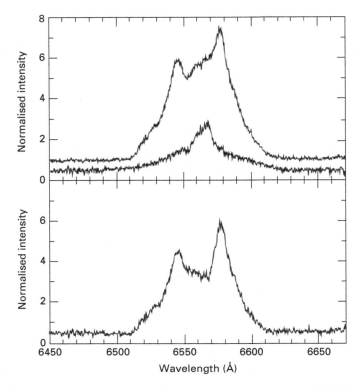

Fig. 7.5 V617 Sgr. Top panel: the 21 May, 2003, Hα profile of V617 Sgr in the high state (upper spectrum), and in early July 2003, in the low state (lower spectrum). Bottom panel: the difference between the high-state spectrum and the decline spectrum. The width of the satellites are ∼600 km s^{-1}(FWHM). In the high state, the width of of Hα was enhanced compared to the low state, with a width of 1650 km s^{-1}(FWHM) (Steiner *et al.*, 2007).

Fast bipolar outflows are suggested from emission-line profiles in U Sco, an eclipsing recurrent nova with a high-mass white dwarf. The ejection velocity reaches 6000 km s^{-1} at the optical peak. Lépine *et al.* (1999) reported a measurement of the acceleration in the outflow, observed five days after the start of an outburst. A spectroscopic time series obtained with the Hubble Space Telescope displayed a bright emission line component moving across the spectrum at a rate of 0.16 ± 0.02 Å in 30 minutes. This is consistent with a clump of emitting material having a line-of-sight acceleration 4.1 ± 0.5 m s^{-2}. The data suggest that this clump resides within a moderately collimated bipolar outflow. Due to the high inclination of the eclipsing binary, this implies an actual acceleration of the clump relative to the star exceeding 25 m s^{-2}.

Kato & Hachisu (2003) noted that the line-of-sight ejection velocity reaches 5000 or 6000 km s^{-1} at the optical peak and gradually decreases in time. Therefore, the jet speed must be nearer 10 000 km s^{-1} given the high inclination angle. Such properties have not been reproduced so far in nova theories. They devised a jet-shaped mass outflow as a mechanism of acceleration. The acceleration and collimation both occur deep inside the region where the spherically symmetric winds would be accelerated in the photosphere just above a convective zone. The jets form within a clumpy medium and are guided in-between the clumps. Thus, continuum radiation is focused and stronger acceleration than in a spherical wind occurs through the super-Eddington flux. In this model, the terminal jet velocity depends predominantly on the white dwarf mass.

Similar high-velocity emission components of H I and He II lines have been observed in other super-soft X-ray sources. For example, RX J0513.9-6951 exhibited emission line components of $3600 \, km \, s^{-1}$ in both positive and negative radial velocities, identified as jets (Crampton *et al.*, 1996).

7.4   Cataclysmic variables

Cataclysmic variables (CVs) are binary systems that consist of a weakly accreting white dwarf and a normal Sun-like star. They are typically small with a separation of roughly the size of the Earth–Moon system and an orbital period in the range 1–10 hours. The companion star loses material onto the white dwarf by accretion. When sufficient material has accumulated and the temperature has risen, a thermonuclear runaway is initiated and a classical nova outburst results. However, although as dwarf nova they show similar outburst behaviour to X-ray binaries, jets have been difficult to detect, which has led to much debate. Even though they contain the best-studied accretion discs, jets were believed to be absent although winds are present. This is now thought merely to represent an observational difficulty (e.g. a lack of material around CVs for the jets to interact with and, hence, to become visible). A second factor is a low accretion rate in comparison with supersoft X-ray sources, which will permit strong radiative cooling and inhibit jet formation in the context of some models (see Section 9.1.2).

One example of twin cataclysmic variable jets is associated with the fast nova V1494 Aql, which was monitored soon after its discovery in December 1999 to September 2000 by Iijima & Esenoglu (2003). The decline and transition in the springtime was accompanied by broad emission wings to the hydrogen lines at high velocity shifts of $-2900 \, km \, s^{-1}$ and $+2830 \, km \, s^{-1}$. This suggests that high-velocity jets were ejected. A second example is the recurrent nova T Pyxidis. Shahbaz *et al.* (1997) presented evidence for a collimated jet in this short-period cataclysmic variable. Optical spectra show bipolar components with velocities $1400 \, km \, s^{-1}$. They argue that a key ingredient for jets is not only an accretion disc threaded by a vertical magnetic field, but also the presence of nuclear burning on the surface of the white dwarf.

Synchrotron emission in some CVs also provides evidence for jets. An example is the dwarf nova SS Cyg. Körding *et al.* (2008) present radio observations of SS Cyg, displaying variable flat-spectrum radio emission in outburst. This is best explained as synchrotron emission originating in a transient jet. Both the inferred jet power and the relation to the outburst cycle are analogous to those seen in X-ray binaries, suggesting that the disc/jet coupling mechanism is ubiquitous. Other CVs with suspected synchrotron jets are AE Aqr and V1223 Sgr. (e.g. Harrison *et al.*, 2010).

7.5   Microquasars: XRB jet systems

The accretor in X-ray binaries (XRBs) may be either a black hole or a neutron star. The accreting material is from a companion star and the systems can be further divided according to the mass of the donor companion: high-mass X-ray binaries (HMXB) and low-mass X-ray binaries (LMXB).

X-ray binaries can be well studied throughout a full outburst cycle because the timescale from quiescence to the peak of the outburst and back ranges from weeks to months. One of the main results of the study of black hole X-ray binaries is the establishment of accretion states through which a source moves in a predefined order. These states have associated jet properties and the jets liberate a significant fraction of the accretion power. The states can be well separated on a hardness–intensity diagram. At the beginning of the outburst, the source shows a hard X-ray spectrum and usually shows radio emission originating from a jet (the hard state). The source brightens while staying in the hard state.

It then makes a transition to the soft state, characterised by a soft X-ray spectrum. The transition is typically accompanied by a bright radio flare and transient jets with high Lorentz factors as the black hole X-ray binary leaves the hard state. After this, the core radio emission is quenched in the soft state. During the decay, the source moves back to the hard state albeit at a lower luminosity than the hard-to-soft transition.

Although the nomenclature of neutron star XRB states is different, one can map the neutron star states onto the black hole equivalents. The main difference is that the radio emission is only suppressed by a factor of 10 when the source is in the analogue state of the soft state. The different behaviour may be due to the existence of a boundary layer in neutron star XRBs, which is absent in the black hole case.

The clear implication is that changes in the accretion disc produce changes in the radio jet, and that an excess of high-energy photons may indicate a source capable of producing highly relativistic radio jets. Finally, with a convincing jet found in an X-ray binary like Cygnus X-3, it is beginning to seem that every galactic X-ray source with radio emission turns out, when imaged, to be a relativistic jet.

7.5.1   Low-mass X-ray binaries

The secondary of LMXB systems is a low-mass star which transfers matter by Roche-lobe overflow. It is less massive than the compact object and can be late-type main sequence, a degenerate dwarf (white dwarf) or an evolved star (red giant). The matter is accreted on to a neutron star from a companion via a disc which radiates in the optical, ultraviolet and X-ray regimes.

The low-mass X-ray binary system Scorpius X-1 probably contains a neutron star as the accretor, has an orbital period of 0.787 days, and lies at a distance of ~ 2.8 kpc. Sco X-1 is persistent as a radio as well as an X-ray source, unlike many of the other well-studied X-ray sources which are transients. Fomalont *et al.* (2001) monitored the system in the radio over four years, including a 56-hour continuous VLBI observation in 1999. They detected compact radio components moving in opposite directions from the radio core. The relative motion and apparent fluxes are consistent with average relativistic speeds of $0.45\,c$ at an orientation of $44° \pm 6°$ to the line of sight. The speeds of different pairs of components at different times range between $0.31c$ and $0.57c$. These synchrotron components are generated from ultra-relativistic plasma that has been expelled from working surfaces where the bulk kinetic energy of jets is efficiently converted. Hence lobes form rather than jet clumps. The lobes have a measured minimum size of 1 milliarcsecond or 4×10^8 km, consistent with an electron radiative lifetime of less than 1 hour. Such a short lifetime can be caused by synchrotron losses or by adiabatic expansion of the electrons at the working surface. The radio luminosity in the frame of the source is 2.0×10^{30} erg s^{-1}, corresponding to a brightness temperature of about 10^9 K. For flares in the core to be correlated with lobe flares requires the signal from a central outburst to propagate down the jets with a speed greater than $0.95c$.

Such radio lobes have been found in several other objects including Cyg X-1. In other well-studied Galactic jet objects the radiating material is probably associated with discrete clouds or shock waves which are still an integral part of the flowing jet. Flares may be associated with directed massive outbursts that are confined to a collimated flow, producing strong radio jets.

Circinus X-1 is a driver of low-collimation X-ray outflow which may be caused by jet precession. It is a highly variable, at times very luminous, X-ray binary, with an orbital period of 16.6 days and a most likely distance of 6.5 kpc. It is an accreting neutron star with thermonuclear flashes on the surface. It has an asymmetric, extended radio structure on arcsecond scales (Fender *et al.*, 1998) with collimated structures extending into the surrounding synchrotron nebula on arcminute scales (Stewart *et al.*, 1993). The radiating electrons in the jet appear to be swept into a wake as the binary perhaps runs away from an associated supernova remnant G321.9-0.3.

An X-ray twin-jet of total length 7" was reported by Angelini & White (2003) to straddle the low-mass X-ray binary and black hole candidate 4U 1755-33. The angular size corresponds to jet lengths of 3–8 pc for possible distances of 4–9 kpc. Further observations in 2004 show that the jets found in 2001 are still present in X-rays (Kaaret *et al.*, 2006). However, sensitive radio observations in 2004 failed to detect the jets. Kaaret *et al.* (2006) show that synchrotron radiation is a viable emission mechanism and that thermal bremsstrahlung and inverse Compton emission are unlikely on energetic grounds. In the synchrotron interpretation, the jet power is 4×10^{35} erg s^{-1}, and the radio non-detection requires a spectral index $\alpha > -0.65$, which is quite reasonable. The source was active for at least 23 years, which would just about be sufficient to form the X-ray jet.

7.5.2 High-mass X-ray binaries

A high-mass X-ray binary (HMXB) is a binary star system that is strong in X-rays, and in which the normal stellar component is a massive star: usually an O or B star, a Be star, or a blue supergiant with a vigorous stellar wind. The massive star is very luminous and dominates the emission in optical light, while the compact object is the dominant source of X-rays. In the high state with X-ray luminosities reaching 10^{36} erg s^{-1}, the radio spectrum is dominated by a partially self-absorbed synchrotron spectrum which may extend as far as the infrared.

One of the most famous high-mass X-ray binaries is Cygnus X-1 at a distance of ~2.1 kpc. It was the first identified as a black hole with mass in the range 7–16 $M_\odot$ and a supergiant secondary of spectral type O9.7 with mass between 20 and 33 $M_\odot$ (Stirling *et al.*, 2001). The orbital period is ~5.6 days and the orbital radius is 0.2 AU. Cygnus X-1 drives a bow shock wave into the ambient medium, producing an optical emission line nebula as well as a jet-blown radio lobe with a size of 5 pc (Russell *et al.*, 2007). This is the result of a strong shock wave that develops at the location where the collimated jet of power $10^{36} - 10^{37}$ erg s^{-1} impacts on the interstellar medium. The shock wave speed is of order

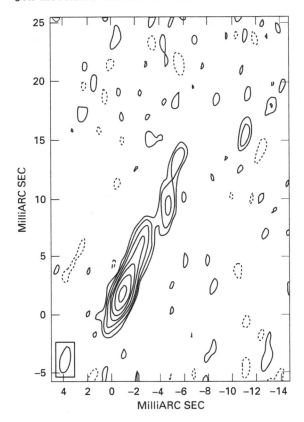

Fig. 7.6 Cygnus X-1. A high-resolution VLBA image of the Cyg X-1 radio jet at 8.4 GHz convolved with a Gaussian beam. The bending at ∼7 milliarcseconds could be a result OF either jet precession or jitter, both of which are observed in SS 433, or due to bending rather than a ballistic motion. The features disappear on a timescale of 2 days. Credit: Stirling *et al.* (2001). Reproduced by permission of the AAS.

$100 \, \mathrm{km \, s^{-1}}$, representing the speed at which the binary moves through the ambient medium. The jet speed is relativistic and most probably in the range 0.3–$0.8c$. It was detected out to 15 milliarcseconds or 32 AU in length and well collimated, with an opening angle of under 2°, as shown in Fig. 7.6 (Stirling *et al.*, 2001).

Cygnus X-3 is thought to comprise a black hole with mass $7 \, M_\odot$ and a companion Wolf–Rayet star at a distance of roughly 10 kpc, with variations interpreted as a 4.8 hour orbital period. Gamma-ray emission is linked to the radio emission from the jet in the system (Tavani *et al.*, 2009). The flux is modulated with the 4.8 hour orbital period, as expected if high-energy electrons are upscattering photons emitted by the Wolf–Rayet star (Fermi LAT Collaboration *et al.*, 2009). In addition, there is a clear time pattern between the γ-ray flares and transitional spectral states of the radio and X-ray emission: in some flares, the particle acceleration occurs a few days before the radio-jet ejections.

Giant radio flares are associated with Cyg X-3. Before a large radio and hard X-ray flare, the soft X-ray is more luminous than usual, and during a flare, the soft X-ray intensity briefly drops. Radio emission is strong and always observable. Radio flares are quickly followed

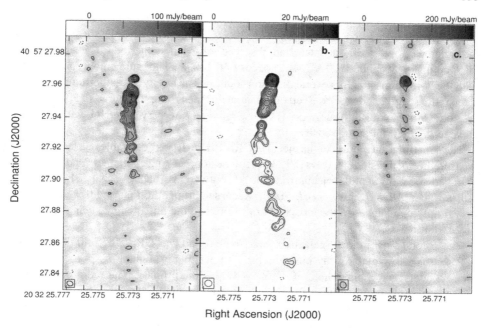

Fig. 7.7 Cygnus X-3. Radio VLBA images of Cygnus X-3, from 1997 February 6, 8, and 11. The jet is at least 50 milliarcseconds long on February 6 and 120 milliarcseconds long by February 8. Credit: Mioduszewski *et al.* (2001). Reproduced by permission of the AAS.

by the ejection of radio-emitting material in the north–south direction with an estimated angular speed of 5 mas d^{-1}.

Mioduszewski *et al.* (2001) detected a curved continuous one-sided jet to the south (Fig. 7.7). The jet curvature changes rapidly, which strongly suggests precession. The ratio of the flux density in the approaching jet to that in the undetected receding jet exceeds 330, implying a jet speed of at least 0.81c if attributed to Doppler boosting. A model for a precessing jet, assuming intrinsic symmetry, yields a jet inclination to the line of sight of under 14°, a cone opening angle under 12° and the precession period must exceed 60 days (Mioduszewski *et al.*, 2001).

7.5.3 *Microquasars*

GRS 1915+105 was discovered by the Compton Gamma Ray Observatory and soon shown to be a hard X-ray transient and then later became the first-known superluminal jet source in the galaxy (Mirabel & Rodríguez, 1994). The black hole mass lies in the range 10–18 $M_\odot$. Flaring events across the spectrum have closely linked activity in the inner accretion disc to the synchrotron jet ejections seen in the infrared and radio. The links are, however, of multiple, apparently distinct, forms. One form releases the large-amplitude superluminal radio events with no infrared jet counterpart.

Superluminal motions have been now associated with several microquasars. Along with GRS 1915+105, the radio jets can be classified into three types: (1) steady plateau-state radio jets, (2) discrete baby jets of 20–40 minute duration in the infrared and radio, and (3) large superluminal radio jets (Dhawan *et al.*, 2000). The steady radio jets are associated with the canonical low-hard X-ray state and observed for extended periods. These are optically thick compact jets under 200 AU with speeds of 0.1–0.4*c*. The radio emission is correlated with the X-ray emission for several different sources. The relativistic superluminal jets have steep radio spectra and are observed out to large distances, a few hundred to 5000 AU, from the core (Mirabel & Rodríguez, 1994). These radio jets are very powerful, with luminosity close to the Eddington luminosity.

The superluminal motion of the ejecta in GRS 1915+105 is detectable in a few hours. The measured velocity is $1.5c \pm 0.1c(D/12 \text{ kpc})$ for the approaching component and is consistent with ballistic motion of the ejecta from 500 AU outward, perhaps even since birth (Mirabel & Rodríguez, 1994; Dhawan *et al.*, 2000). Recorded proper motions and flux density ratios vary, implying changes in the jet speed or the angle to the line of sight such as effected by precession.

The second superluminal galactic source to be found was GRS J1655-40 (Tingay *et al.*, 1995). It is a low-mass X-ray binary composed of a blue subgiant of spectral type F4 as the secondary and a black hole of mass $\sim 7 M_\odot$ as the primary. The radio images reveal two components moving away from each other at an angular speed of 65 ± 5 mas d^{-1}, corresponding to superluminal motion at the estimated distance of 3–5 kpc. The 12-day delay between the X-ray and radio outbursts suggests that the ejection of material at relativistic speeds occurs during a stable phase of accretion onto a black hole, which follows an unstable phase with a high accretion rate.

7.5.4   *SS 433*

One of the most exotic star systems observed is the eclipsing X-ray binary system SS 433, at a distance of ~ 5.5 kpc (Margon, 1984). The designation derives from the San-duleak and Stephenson 1977 catalogue of stars with strong Hα emission lines, being the 433rd entry. The primary is almost certainly a black hole rather than a neutron star (Blundell *et al.*, 2008). With strong radio jets, SS 433 is classified as a microquasar, the first discovered. The secondary companion star is suggested to be a late A-type star. The primary and secondary orbit each other at a very close distance in stellar terms, with an orbital period of 13.1 days.

The mysterious SS 433 jets produce optical spectral lines even though they are relativistic. Remarkably, the cool jet material travels at 26% of the speed of light. Therefore, for once, we know not only the jet speed but also the jet composition, which must include baryons rather than just electrons and positrons. The jets from the primary are emitted perpendicular to its accretion disc, above and below the plane. The jets and disc precess around an axis inclined about 79° to our line of sight. The angle between the jets and the axis is $\sim 20°$, and the precessional period is around 162.5 days. The precession means that the jets corkscrew through space in an expanding helical spray, producing both blue and red Doppler shifts in the optical (Margon *et al.*, 1979) and X-ray emission lines. The model defined by the emission lines successfully predicts the changing pattern of the radio emission, as displayed in Fig. 7.8. This confirms the ballistic motion of the radio structure. The radio structure consists of clumps which brighten soon after ejection. On the large scale, the jets

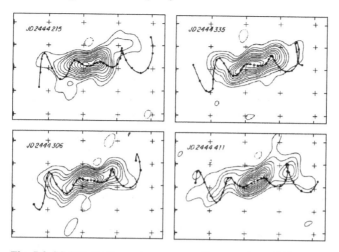

Fig. 7.8 SS 433. Radio contour maps at four epochs at 5 GHz acquired with the VLA. An unresolved radio core has been removed. The precession model indicates 20-day intervals in ejection events. The + signs indicate 1″ intervals in declination and 0.1 second in right ascension. Credit: Hjellming & Johnston (1981). Reproduced by permission of the AAS.

impinge the supernova remnant W50 within which SS 433 lies centrally, distorting it into an elongated shape.

The ejected radio components brighten as they are released. The material the jets are impacting appears to be replaced some of the time, leading to variations in the brightening of the jets. Variations in jet speed as well as the angles, lasting for as long as tens of days, are necessary to match the detailed structure of each jet. It is remarkable that these variations are equal and opposite, matching the two jets simultaneously (Blundell & Bowler, 2004). The magnetic field in the jets is aligned parallel to the local velocity vector, strengthening the case for the jets to be composed of discrete bullets rather than being continuous flux tubes.

The spectrum of SS 433 is determined not just by Doppler shifts but also by relativity. When the effects of the Doppler shift are subtracted, there is a residual redshift which corresponds to a velocity of about 12 000 km s^{-1}. This is due to pure time dilation as the relativistically moving excited atoms in the jets appear to vibrate more slowly and their radiation thus appears redshifted.

SS 433 is regarded as a super-Eddington accretor in which the accretion state is near or above the Eddington luminosity (at which the outward radiation force overcomes gravity). The supercritical rate of accretion on to the central black hole is reduced, regulating the luminosity to the Eddington limit. Super-Eddington accretion discs are generally expected to possess vortex funnels and radiation pressure-driven jets from geometrically thick discs However, there is still much speculation because of the rarity of such sources. The accretion disc itself is subject to extreme heating and produces intense X-rays. The X-ray reheating mechanism is transient and does not correlate with the total flux in the core or in the extended radio jets.

SS 433 may be dwarfed by a more distant cousin. The most powerful super-Eddington accretor with microquasar jets may be contained in a nearby galaxy. The large nebula S 26 in the galaxy NGC 7793 is powered by a black hole with a pair of collimated jets (Pakull *et al.*, 2010). The jets are inferred from Chandra X-ray images, which resolve three perfectly aligned components that match the 300 pc extent of the major axis of the optical nebula. The interpretation is that the outer components are hot spots where the jets impact the ambient medium, analogous to a Fanaroff–Riley Type II radio galaxy. The mechanical power is estimated to be a few 10^{40} erg s^{-1}. The jets are then 10^4 times more energetic than the X-ray emission from the core. It may thus be a microquasar where most of the jet power is dissipated in thermal particles rather than relativistic electrons.

Microquasar jet properties have been hard to pin down. Interpretation of the X-ray emission from SS 433 has made it the exception. In addition to thermal bremsstrahlung, strong iron lines are found which shift as expected from the jet optical lines. The X-rays originate from just 10^{12} cm of the compact object, whereas the optical lines are produced much further out at distances up to 10^{15} cm (Watson *et al.*, 1986). A density of $\sim 10^{13}$ cm^{-3} on the scale of 10^{12} cm, temperature 3×10^7 K and X-ray luminosity of 10^{36} erg s^{-1} lead to a mass outflow rate of $1.6 \times 10^{-6} f \, M_\odot$ yr^{-1} where f is the fraction of the jet volume occupied by the hot gas.

7.6   Pulsar jets

Pulsar wind nebulae are associated with supernova remnants in some cases where a pulsar is still inside its associated supernova remnant. These strong X-ray-emitting regions may be powered by the pulsar's relativistic wind and confined by the high-pressure supernova remnant environment. Besides the wind, jets are occasionally present.

X-ray jets have been detected inside the supernova remnant MSH 15-52 driven by the young, energetic 150 ms pulsar PSR B1509-58. The radio appearance is dominated by two spots, a brighter one to the north-west and a fainter one to the south-east. The radio emission to the north-west coincides spatially with the H II region RCW 89. The Chandra satellite revealed two X-ray jets in the south-east and north-west directions, the latter terminating in the optical nebula RCW 89 (Gaensler *et al.*, 2002). The main collimated feature is over 4' long and $\sim 10"$ wide, lying along the nebula's main symmetry axis, which may well correspond to the pulsar spin axis. A speed exceeding $0.2c$ is suggested on cooling time grounds. The lack of an observed counterjet implies that the pulsar spin axis is inclined at $\sim 30°$. The second jet is an elongated region of reduced emission which does not begin until 3' from the pulsar and then extends along the nebular axis.

The Crab pulsar drives a faint, filamentary 'chimney' or 'jet' of emission filaments, first detected in the optical by van den Bergh (1970). This is probably a trail of shocked gas and not a bona fide jet. However, there are physically much smaller X-ray synchrotron jets on a scale of 0.1 pc, emanating directly from the central pulsar, shown in Fig. 7.9. In X-rays, the pulsar is centred on a torus with the suggestion of a hollow-tube structure (Aschenbach & Brinkmann, 1975). This morphology has been confirmed and a jet and counterjet detected nearly along the axis of the torus (Aschenbach & Brinkmann, 1975; Weisskopf *et al.*, 2000).

A similar X-ray jet and ring-like structure is present for PSR B0833-45, the 89-millisecond pulsar associated with the Vela supernova remnant (Fig. 7.9). The physical size of the Vela wind nebula is only 0.1 pc since it is close by at 250 pc. A linear, jet-like feature of 10", or a deprojected length of 4.1×10^{16} cm, which emanates from the pulsar towards the south-east (e.g. Helfand *et al.*, 2001) can be interpreted as synchrotron radiation.

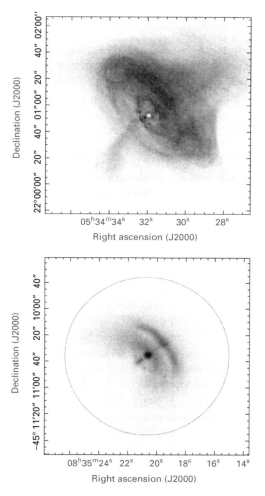

Fig. 7.9 Pulsar jets. The X-ray jets and relativistic wind nebulae surrounding two young pulsars observed by the Chandra Observatory: the 1000 yr Crab pulsar (top) and the 10 kyr Vela pulsar (bottom). The Vela 'Dragonfly' nebula is 16 times smaller physically, assuming distances of 2 kpc (Crab) and 250 pc (Vela); the circles in the two images represent the same physical size. Credit: Helfand *et al.* (2001). Reproduced by permission of the AAS.

There is also evidence for a counterjet. As shown in Fig. 7.9, the jets have a position angle of 130° and are aligned to within ∼ 8° of the pulsar's proper motion vector. The jet cannot simply be a wake, but must be continuously supplied with particles from the pulsar. The luminosity required is only 5×10^{30} erg s^{-1}, roughly 1% of the nebular luminosity.

A faint, curved outer jet extends approximately in the same direction as the Vela jet out to ∼100" in approximately the same direction (Kargaltsev *et al.*, 2003). This outer jet is comparable in luminosity to the inner jet but strongly variable in shape and brightness. Bright blobs recede from the pulsar with apparent speeds in the range 0.3–0.6*c*, fading over days to weeks. Kargaltsev *et al.* (2003) suggest that the outer jet supplies an extended region of diffuse emission.

Accretion discs are not associated with pulsars such as the Crab and Vela even though jets are detected. However, Blackman & Perna (2004) demonstrate that an accretion disc radiating below detectable thresholds may still account for much of the pulsar activity, including the jets. Nevertheless, little is known concerning their launch.

7.7   Gamma-ray bursts

Jets are intimately associated with gamma-ray bursts (GRBs). Although one jet is detected, twin jets should be launched by the central engine according to all popular progenitor models. Rather than outstanding examples, the observations are best described in terms of common behaviour.

Gamma-ray bursts are detected roughly once per day from entirely random directions of the sky. Associated with catastrophic events, they do not repeat. They are divided into two classes based on the duration and hardness of their prompt emission spectra. Long-duration bursts (LGRBs) last anywhere from 2 seconds to a few hundreds of seconds. These were the first class for which afterglow observations were possible, which showed that they were cosmological, occurring in star-forming galaxies at high redshifts (Costa *et al.*, 1997). We now know that GRBs are associated with the deaths of massive compact progenitor stars (Woosley, 1993), take place in compact star-forming galaxies and have been detected up to a redshift of $z = 8.2$ (Tanvir *et al.*, 2009; Salvaterra *et al.*, 2009).

With the detection of GRB 030329, the association of long-duration GRBs with super-novae was firmly established, which strongly supports the collapsar model as the mechanism behind long-duration GRBs (Hjorth *et al.*, 2003). Rapid follow-up observations to the BATSE discovery measured an optical afterglow of magnitude ~ 12 in the R-band within 1.5 hours, brighter than any previously detected afterglow at a comparable time after burst. A very bright afterglow was also detected from radio to X-rays (Fig. 7.10). The isotropic calculated total energy release was $\sim 9 \times 10^{51}$ erg, and thus a classical cosmological GRB.

The light curves for GRB 030329 in the optical, X-ray, millimetre and radio are displayed in Fig. 7.10. The optical light curve reveals a rich array of variations, superposed over the mean power-law decay (Lipkin *et al.*, 2004). The variations maintain a similar timescale during the first 4 days and then get significantly longer. The structure of these variations is similar to those previously detected in the afterglows of several GRBs. A break in the optical decline, as expected in relativistic jet models, takes place between 3 and 8 days after the burst, as shown in Fig. 7.11.

Through a process of elimination, Berger *et al.* (2003) settled on a two-component explosion model for GRS 030329 in which the first component was a narrow jet of 5° with an initially large Lorentz factor. It is responsible for the gamma-ray burst and the early optical and X-ray afterglow, including a break at 0.55 days. The second component is a wider jet of 17° which powers the radio afterglow and late optical emission. The break due to the second component is readily seen in the radio afterglow but is masked by SN2003dh in the optical bands. Such a two-component jet finds a natural explanation in terms of the collapsar model, as some material briefly involved in the explosion, rather than the collapse to the black hole, abjectly fails to escape and collapses towards the black hole (MacFadyen *et al.*, 2001).

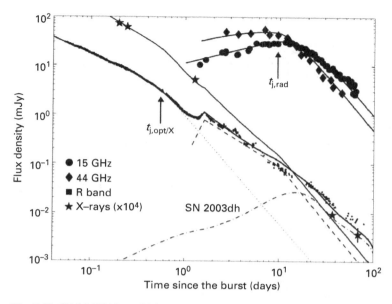

Fig. 7.10 GRS 030329: multi-band light curves. Data are shown as symbols, as labelled, and the solid curves are the model predictions for an updated two-jet model in each respective band. The radio emission is suppressed by synchrotron self-absorption at early times. The underlying light curve of the supernova SN 2003dh is also shown. Credit: Lipkin *et al.* (2004). Reproduced by permission of the AAS.

The association of long-duration gamma-ray bursts with Type Ibc supernovae presents a challenge to supernova explosion models. In the collapsar model, bursts are produced in an ultra-relativistic jet. The jet is launched from the magnetosphere of the black hole that forms in the aftermath of the collapse of a rotating progenitor. The jet is powered by a continuous infall and disc-like accretion of the progenitor's interior and driven along the star's rotation axis. While the collapsar model seems to successfully explain the power and duration of LGRBs, it is not clear at present whether it naturally gives rise to a supernova-like stellar explosion.

There is now compelling evidence of a link between long-duration gamma-ray bursts and Type Ib/c supernovae. The lack of hydrogen lines in spectra is consistent with model expectations that the stellar progenitor lost its hydrogen envelope to become a Wolf–Rayet star before exploding. These results have given a definite impetus to the collapsar model in which a Wolf–Rayet progenitor undergoes core collapse, producing a rapidly rotating black hole surrounded by an accretion disc, which injects energy into the system and thus acts as a central engine. The energy extracted from this system supports a quasi-spherical Type Ibc supernova explosion and drives collimated jets through the stellar rotation axis, which produce the prompt γ-ray and afterglow emission

Many aspects of the observed long-duration GRB behaviour can be explained by the relativistic fireball model (Starling *et al.*, 2009). The prompt emission is largely produced by shock waves internal to the outflow. The long-lived afterglow is also produced by shock waves as the outflow impacts against the ambient medium. However, as fireballs, they emit

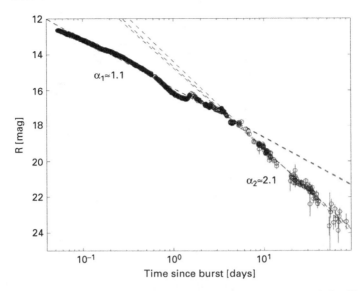

Fig. 7.11 GRS 030329 afterglow. This R-band light curve was derived by subtracting the contributions of the supernova SN 2003dh and the host galaxy from the observed light curve. Overlaid are power-law fits to early and late times. Credit: Lipkin *et al.* (2004). Reproduced by permission of the AAS.

isotropically and the radiative energies implied for some bursts would be extremely large. Since no known mechanism can produce high-energy photons with efficiency approaching 100%, the total explosive energy required would be even greater, implausibly large for a stellar core-collapse powered event.

The energetics argument led to the expectation that the outflow must be confined to a jet, reducing the overall energy requirements (e.g. Rhoads, 1997). In that case, a sufficiently massive core collapsing to a black hole might produce an outflow which could pierce a hydrogen-stripped envelope to still produce a relativistic jet. This picture received support from the observation of concurrent supernova events at GRB sites for some low-luminosity bursts (e.g. Hjorth *et al.*, 2003). The hydrodynamics of the jet will be the same as that of a fireball provided the Lorentz factor Γ exceeds $1/\theta_o$, where θ_o is the jet opening angle. The matter then does not have time to expand laterally. The jet may be considered to expand at the local sound speed which, for an ultra-relativistic plasma yields an opening angle of $\theta \sim \theta_o + c_s t_{proper}$, where t_{proper} is the proper time. This yields $\theta \sim \theta_o + 1/(\Gamma\sqrt{3})$. The expansion can, however, be with a Lorentz factor equal to the thermal Lorentz factor, which yields $\theta \sim \theta_o + 1/\Gamma$ (Sari *et al.*, 1999).

Theoretically, the collapse of a massive star will almost inevitably lead to the birth of a black hole surrounded by an accretion disc. It is logical that the symmetric accretion system will produce double-sided jets. The GRB, however, can only be observed when our line of sight falls close to the axis of one of the two jets. A strong prediction of all jet scenarios is that the observed light curve should exhibit an achromatic break, visible in both optical and X-ray light curves, when the relativistic outflow slows to the point that the Doppler beaming angle becomes wider than the opening angle of the jet: the later the break time, the wider the jet (Sari *et al.*, 1999). However, in many GRBs no temporal break is seen at all to late times, implying much more energy than expected. The most extreme example is that of GRB 060729, which continued a smooth X-ray power-law decline for 125 days, implying an opening angle greater than 28° (Grupe *et al.*, 2007). The jet flow models can also be tested observationally through other relativistic beaming effects, including the polarisation properties in both the prompt and the afterglow phase and the predicted existence of orphan afterglows (i.e. an afterglow not preceded by prompt γ-ray emission due to the viewing angle).

Observational evidence has accumulated supporting the idea that many GRBs are highly collimated. Frail *et al.* (2001) showed that the energy release is then narrowly clustered around 5×10^{50} ergs after taking into account conical jet structures. After correction for collimation, nearly all bursts access a rather standard energy reservoir which is of a comparable energy to that of ordinary supernovae. In addition, the broad distribution in fluence and luminosity for GRBs is largely the result of a wide variation of opening angles. Thirdly, the true GRB rate is several hundred times larger than the observed rate, with only a small fraction of these visible to a given observer.

Much less is known about the class of short-duration gamma-ray bursts and their progenitors. It has been suspected for a long time that SGRB progenitors are binary systems of two compact objects that merge after their orbital energy has been radiated, mostly in the form of gravitational waves. The localisation of short-duration GRBs has confirmed that some are associated with early-type galaxies and explode at relatively large distances from the centre of the host galaxy, Most of them closely follow star formation and cannot therefore originate from a traditional merger path. This also suggests that a large fraction are also associated with massive stars. A close association between the progenitors of SGRBs and LGRBs was theoretically explored within the multi-jet scenario in which different viewing angles of the same set of GRBs determine the GRB properties (Yamazaki *et al.*, 2004; Toma *et al.*, 2005). Long-duration GRB progenitors may well produce short-duration GRBs at large off-axis angles, 40°–50° away from the axis along which long-duration GRB ejecta are released (Lazzati *et al.*, 2010).

7.8   Summary

Jets are generated under an amazing variety of conditions but not in all circumstances. No general pattern has emerged and we are set to continue to try to understand the configurations which lead to specific jet-producing objects. Stars known to be members of binary star systems are prone to jet formation but it is not clear how much a combination of substantial accretion and magnetic field is necessary or sufficient. However, new developments are keeping the research field vibrant. For example, transient jets in X-ray binaries appear just as an accretion disc undergoes a transition in states, suddenly losing the inner annuli. In addition, relativistic jet physics has found a new unexpected application of immense importance as the key to interpreting the new phenomena associated with gamma-ray bursts.

8

Jets within the solar system

We have learnt about nearby jets by going out to meet comets as they come in to visit our neighbourhood. Jets of gas and dust appear as comets approach the Sun. Since Vega and Giotto's rendezvous with Halley in 1986, we have known that the jets are generated from within the nucleus of a comet, and that the nucleus is only a few misshapen kilometres of low-density ice and rock. Such jets change the trajectory of a comet, providing a rocket effect, which alters the comet orbit. In addition, mass loss through jets produces the spiral and shell structures seen in the comae and proceed to supply the tails. The mass loss also implies that several metres depth of material is lost from an active region per orbit around the Sun.

Our capability for close encounters has resulted in the only resolved images of the launching regions of astrophysical jets. Here, we first synthesise the observations of cometary jets and establish their properties. In Chapter 9, we will see how the probable launch mechanism compares with other models.

Solar jets have also come into prominence in the last decade through high-resolution satellite imaging. On all scales, from spicules to coronal jets, the Sun displays a bewildering array of jet-like phenomena. The uniting factor is that they are magnetically rooted in the photosphere. Some propagate into the chromosphere, while others shoot up into the corona. Apart from exploiting the same energy source, they really have nothing to do with cometary jets. Rather, being magnetically mediated, the physics is relevant to more distant jets although the *obligatory* accretion disc is missing.

8.1   Cometary jets, pre-2000

Jet-like gas and dust outflows have long been observed from many comets besides Halley. The dust is usually observed in the optical through scattered light from grains smaller than 1 μm. One exception was the detection of HCN gas at millimetre wavelengths from Comet Kohoutek (1973f) before and after perihelion passage (Huebner *et al.*, 1974). For Kohoutek, jet speeds ranging up to several kilometres per second were deduced from multiple Doppler shifts in the observed spectrum (Huebner *et al.*, 1974).

In May 1996, Comet Hyakutake reached its closest point to the Sun, well inside the orbit of Mercury. Inner dust jets were then seen to transform into arcs around the nucleus, resembling a pinwheel. This effect is due to the rotation of the comet's nucleus. Figure 8.1 shows two jets and two expanding arcs. The inner arc is about 8000 km and the outermost arc is at a projected distance of 12 000 km from the nucleus. The arcs expand away from the

Fig. 8.1 Comet Hyaktake. Optical continuum image of Comet Hyakutake acquired with the USNO's 24-inch Telescope. Credit: James A DeYoung (USNO).

nucleus at 0.25 km s^{-1} while the nucleus rotates with a 6.3-hour period. The brightest of the Comet Hyakutake's jets attaches to the inner arc. Schleicher & Woodney (2003) reported an average dust outflow velocity of 0.38 km s^{-1} between 3000 km and 8000 km.

The large-scale Hale-Bopp jets shown in Fig. 1.1 display some curvature as a result of the 11.75-hour rotation period of the nucleus and a dust speed of 0.41 km s^{-1} measured over the scales of 5000 km to 45000 km (Warell *et al.*, 1999). The jet behaviour resembled sprinkler heads as the nucleus rotated (Licandro *et al.*, 1998). The total mass outflow rate from Hale-Bopp reached extremely high levels with a remarkable dust outflow rate of 1.4×10^5 kg s^{-1} (Lisse, 2002). The high jet activity near perihelion supports the idea that jets might dredge up some specific identified grains from deep within the nucleus, rather than being lifted off the surface of the comet, and that the size of pores on the porous surface may increase as the distance from the Sun decreases (Harker *et al.*, 2002). These pores may, however, be only 100 μm wide and so may be easily clogged up by large grains. It should also be noted that gas jets were directly observed in the inner Hale-Bopp coma. However, most of the HCN gas detected at 88 GHz is roughly spherically distributed (Veal *et al.*, 2000). Hence, especially for Hale-Bopp, any jet model must be able to account for either a loss in collimation or a dominant spherical component.

Halley dust jets were observed in detail via scattered light from the ground and space in 1986. The dust emission from near the nucleus is associated with only 10–15% of the surface area which is generally extremely dark (Keller *et al.*, 1987). Thus Halley is described as a snowy dirtball rather than a dirty snowball with volatiles (80% H_2O) contained under the surface. The strongest activity is associated with the afternoon side of the nucleus (Fig. 8.2). Three prominent jets and many small jet-like features were observed with individual active regions typically 10 km^2 in area. Narrow sub-jets with full opening angles of 10° and width 0.5 km were identified by Keller *et al.* (1987). The jets were not radial but crossed each other at angles of up to $\sim 20°$.

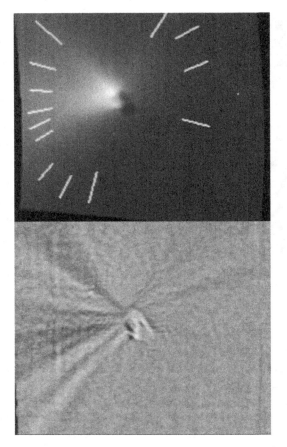

Fig. 8.2 Comet Halley. *Upper panel*: Locations of 13 jets from the near-nucleus region of Comet Halley recorded by the Halley Multicolor Camera on Giotto. The nucleus itself is roughly elliptical with size 15 km × 8 km. Lower panel: An azimuthal gradient technique has been applied to enhance the structure (Keller *et al.*, 1987).

8.2   Cometary jets, post-2000

Since Halley there have been four further spacecraft fly-bys, all to Jupiter-family members: 19P/Borrelly (Deep Space 1), 81P/Wild 2 (Stardust), 9P/Tempel 1 (Deep Impact) and 103P/Hartley 2 (EPOXI mission, Deep Impact)).

Deep Space 1 passed 2200 km from Comet Borrelly in 2001 and obtained an image resolution of 63 m. The surface of the comet, and probably most active comets, is dominated by features directly or indirectly formed by sublimation of volatiles. The nucleus is very dark with albedo variations from 0.007 to 0.035, with less than about 10% of the surface actively sublimating (Boice *et al.* 2002b). Ballistic morphologies are either absent or rare (Boice *et al.* 2002b).

A particularly bright jet appeared sharp and well collimated, as shown in Fig. 8.3 and Fig. 8.4. It remained fixed in orientation, perpendicular to the 8 km long axis of the flattened nucleus, and consistent with a location close to the rotation axis, being the minor

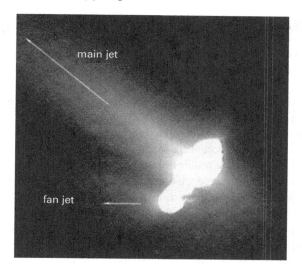

Fig. 8.3 Comet Borrelly image taken by Deep Space 1 on 22 September 2001 from a distance of 4825 km at a resolution of 63 m per pixel, enhanced to reveal dust jets. The main dust jet is directed towards the top left of the frame, around 35° away from the comet–Sun line (the Sun being to the left of the page). The jet base consists of at least three smaller sub-jets. Credit: NASA/JPL, Yelle *et al.* (2004).

Fig. 8.4 Comet Borrelly. A well-collimated sunward-directed dust jet from a Comet Borrelly image, enhanced to reveal dust jets by Farnham (2009). The jet with opening angle of 20–30° has been enhanced by division of a $1/R$ profile.

principal axis (Soderblom *et al.*, 2004). The jet is actually resolved into smaller sub-jets, each appearing cylindrical with radii 200–400 m separated by about a kilometre. Two of these columns are traced down to dark patches adjacent to bright terrain (Yelle *et al.*, 2004). This terrain is described as a smooth, broad basin containing brighter regions and mesa-like structures.

The intensity of the Borrelly dust jets displays a $1/R$ dependence at distances, R, larger than about 20 km from the nucleus (Boice *et al.* 2002a). As the optical depth of the dust near the nucleus is small ($\tau \sim 10^{-3}$), this is interpreted to also be the distribution of the dust column density. This behaviour is expected for free dust outflow.

Near the nucleus, however, the intensity profiles of the major jets and sub-jets reveal a strong flattening relative to the inverse-distance relationship. Similar profiles were observed in the inner coma of Halley from the Giotto spacecraft, which were modelled by Huebner *et al.* (1988). For Halley, an intensity profile at large distances from the nucleus appears as if originating from a virtual point below the surface. The profile flattening is then interpreted to be a result of the extended size and nonuniformity of the active region.

Stardust flew to within 237 km of Wild 2 on 2 January, 2004. Stardust detected 20 jets, mainly on the sunlit side of the comet but, surprisingly, two on the dark side. They contain water, carbon dioxide and rock particles. Sixteen jets emanate from sources that are on slopes where the Sun's elevation is greater than predicted from fitted triaxial ellipsoids (Sekanina *et al.*, 2004). A few major jets originated from an area within 25° of the spin axis and were in continual sunlight. Many of the jets were found to emanate from 'pit-halo' regions on the nucleus of the comet that the science team named *Mayo* and *Walker*. None of the jets emerged from the flat-floored *Left Foot* and *Right Foot* features close to Wild 2's sunlit pole.

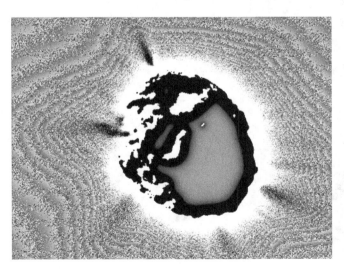

Fig. 8.5 Comet Wild. A view of both the 4-km nucleus and the jets of Comet Wild 2 is made possible by the superposition of a long- and a short-exposure image. The two images were taken 10 seconds apart from a distance of about 275 km just seconds after Stardust's closest approach to Wild 2. Streams of dust can be traced back to lumpy pits on the surface of the comet's nucleus. Credit: NASA/JPL

Also in 2004, carbon dioxide and cyanogen gas jets were observed from the ground, spiralling off Comet Machholz (Lin *et al.*, 2007; Farnham *et al.*, 2007). The jet source locations were found to be on opposite hemispheres at mid-latitudes. The jets are oppositely directed, at longitudes about 180° apart. The CN features were measured to be moving at about 0.8 km s^{-1}.

On 4 July, 2005, the Deep Impact spacecraft (a 370 kg copper projectile) crashed into Comet Tempel 1 at a speed of 10.2 km s^{-1}. During the approach phase, jets were observed in the coma, turning on and off and producing arcs (archimedean spirals) that expanded outward from the rotating nucleus. Three distinct jets rotating with a 1.7-day periodicity were followed. The brightest jet produces an arcuate feature that expanded outward with a projected velocity of about 12 m s^{-1}. During the close approach, linear jets thought to be the inner segments of the archimedean spirals were found (Fig. 8.6). Because of the 41-hour rotation period and the rapid fly-by, changes in these jets are predominantly due to perspective differences in the spacecraft's viewpoint. Thus, pairs of images contain stereo information that can be used to trace the jets back to their origins on the nucleus. In fact, through a remarkable sequence of images, the origin of jets has indeed been traced back to the surface, providing the most direct evidence for jet formation (Fig. 8.7).

Adjacent to bright ice patches are small, dense dust jets that appear to project directly from the surface (Farnham *et al.*, 2007). This image strongly suggests that the jets are produced from small, active vents of size ~ 10 m, driven by the sublimation of sub-surface ices. In addition, the association of CO_2 emission indicates that the vents remained active well after sunset of the previous rotation, which would aid in the ejection of water that recondenses on the immediately surrounding surface. Therefore, the ice patches, which should be short-lived, can be periodically replenished.

In addition, a jet may be just strong enough to keep itself from choking and so deliver some of the ejected dust material back onto the mantle, helping to form the inactive regolith

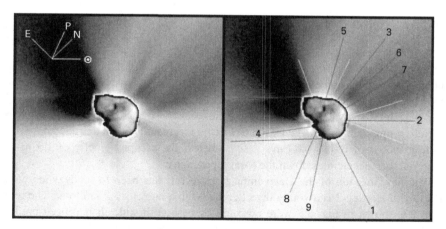

Fig. 8.6 Comet Tempel viewed from Deep Impact at a range of 1.4×10^4 km and a field of view of about 28 km. The image is enhanced to improve the contrast of the nucleus and coma. The left panel shows the enhanced image, while the right panel shows the same image with radial jet features highlighted. The sources of the jets may lie on the far side or near side. Credit: Farnham *et al.* (2007)

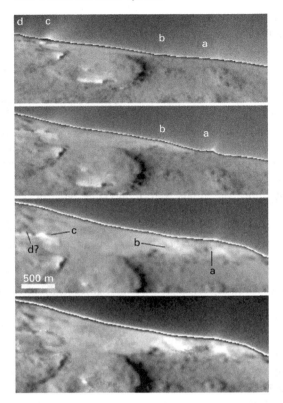

Fig. 8.7 Comet Tempel. Deep Impact images showing the small surface jets on the limb of Comet 9P/Tempel at the northern limb obtained over a period of 80 s (Farnham *et al.*, 2007). As the spacecraft passed under the nucleus, the jets appeared to cross over the horizon. The approximate sources of the jets are marked in the third panel.

elsewhere (Bar-Nun *et al.*, 2008). With flow speeds of just 1 m s^{-1}, material may reach heights of a few kilometres before falling back under gravity, forming structures similar to those of volcanic eruptions. This could also produce the smooth terrains on Tempel's surface (Belton & Melosh, 2009). Due to the weak tensile strength, crystalline ice becomes locally ruptured and fluidized by the CO gas pressure and is then extruded onto the surface at low speeds of 0.003–0.03 m s^{-1}, well below the escape velocity of 1.3 m s^{-1}. This material then accelerates down the local topography.

The new mission of the Deep Impact spacecraft has been to rendezvous with Comet Hartley 2. Hartley 2 is a young dwarf comet with a nucleus 2.2 km long. The remarkable dust jets shown in Fig. 8.8 were powered by carbon dioxide. It had been predicted that sublimation in Hartley would not be controlled by water ice but by CO or CO_2 (Ferrín, 2010), with CO_2 earlier detected spectroscopically. Water vapour is also detected but not from this part of the comet, and water ice has been detected in the form of chunks or snowballs. The jets were easily detected, not requiring detailed processing to enhance narrow features such as was necessary for Comet Tempel.

Fig. 8.8 Comet Hartley. A colour-reversed image of Comet Hartley 2 taken in November 2010 from Deep Impact at a distance of about 700 km. The Sun illuminates the nucleus from the right. Image credit: NASA/JPL-Caltech/UMD.

In summary:

(1) Jets can be stable and long-lived, lasting for months. However, they can also be extremely dynamic, changing their appearance or direction on timescales as short as hours.
(2) Jets have been detected individually and in groups, and have no preferred orientation. Although both fans and jets are primarily radial, they can both show curvature at increasing distances from the nucleus.
(3) To gain an estimate of the surface area involved in sublimation, measurements of the water production rate are deduced from the proxy OH molecule. For most but not all comets, the active fraction of the surface is only a few per cent.
(4) Quantitatively, very little has been derived directly from the observations of individual dust jets and the supersonic nature of driving gas jets must often be deduced from the dust component.

Indeed, the collimation for the gas is more of an issue than for the dust. For example, even though it has a highly collimated dust jet, Comet 19P/Borrelly shows little collimation of the gas, at least on large spatial scales. This seems to support the conjecture that both the gas and dust are originally collimated but decouple very close to the nucleus, allowing the dust to continue vertically while the gas then expands laterally.

8.3 Moon jets

Jets of gas and dust were discovered emanating from the south polar terrain of Enceladus by the Cassini spacecraft (Porco *et al.*, 2006). Enceladus is Saturn's sixth largest moon, a 500-km body with a mean density of 1.6 g cm^{-3} and an icy surface layer possibly

Fig. 8.9 Enceladus. Reversed-colour Cassini image of the ice particle jets on the south pole of Enceladus. (Credit: NASA/JPL/Space Science Institute)

of depth 10 km. The south polar region is anomalously warm with four tiger-stripe fractures that harbour the base of the jets. The jets are local vents situated within the stripes.

The jets, visible through the dust component in Fig. 8.9, contribute on a larger scale to giant plumes. The plumes reach widths of 10 km at an altitude of 15 km but extend hundreds of kilometres above the surface. The total mass ejection rate is estimated to lie in the range $2-16 \times 10^5$ g s^{-1} or $10^{28}-10^{29}$ mol s^{-1} (Saur *et al.*, 2008), which would be sufficient to supply oxygen to the system and, even with some loss, replenish the icy particles of Saturn's E ring. The large mass variations were measured between fly-bys separated by months; the actual variability time scale may be much smaller.

The jets are dominated by water vapour with a temperature of ~ 150 K, containing not just H_2O (91±3%) but with N_2 (4%), CO_2 (3%) and CH_4 (2%) detected. The gas moves at a mildly supersonic speed of ~ 0.3 km s^{-1} (Hansen *et al.*, 2008) while the escape speed from the moon is ~ 0.24 km s^{-1}, so that most of the gas escapes. On the other hand, micrometre-sized dust achieves typical speeds in the range 80–160 m s^{-1} (Hedman *et al.*, 2009). The dust is dominated by fine-grained ice particles (Teolis *et al.*, 2010). Therefore, most of the dust in the jets probably returns to the surface, which is found to be covered by particles typically tens of micrometres in size, as well as blocks of ice tens of metres in size which were probably also ejected through vents (possibly explosively, driven by gas in cryovolcanic events).

Jets have also been invoked from observed surface features on Mars (Kieffer H. H. *et al.*, 2006). Evidence for venting and the existence of plumes suggests that jets might occur on other solar system bodies. However, volcanic ejection of magma or ice particles does not necessarily imply the combination of supersonic speeds, sufficient collimation and sustained ejection necessary to be designated as a jet.

8.4   Solar jets

Jet-like events are very common in the solar atmosphere (Pariat *et al.*, 2010). Recent observations with Hinode and Yohkoh have revealed that coronal jets are a much more frequent phenomenon than previously believed. They usually possess good collimation and extend radially. Some jet events can be observed at multiple wavelengths. The following list emphasises the broad range of wavelengths and the vast range of scales on which they are now observed.

1. Spicules and mottles (e.g., Tziotziou *et al.* (2003); De Pontieu *et al.* (2007b)). The chromosphere, especially the upper parts, consists almost entirely of elongated thin jet structures known as spicules, with lifetimes of approximately 10 minutes (Fig. 8.10). Hundreds of thousands exist on the Sun at any given time. Similar features seen in quiet regions of the solar disc are called mottles. They are considered to be the disc counterparts to the limb spicules and also the principal channels through which mass and energy are supplied

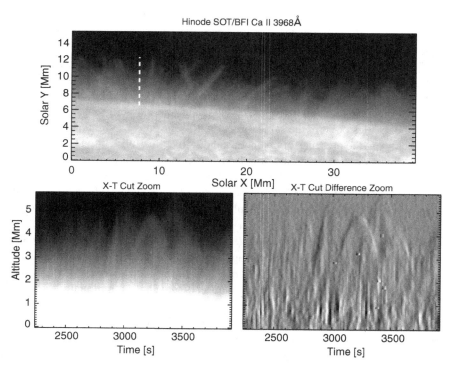

Fig. 8.10 Spicules on the limb of the quiet Sun imaged by Hinode on 2006 November 22, with the SOT in the Ca II H 3968Å line (top panel). The lower panels display space–time plots along the location indicated by a dashed line in the top panel. The short-lived vertical stripes (type II spicules) and longer-lived parabolic paths (Type I spicules) are two different populations governed by very different timescales. To enhance these types, Fourier filtered time-series were created to isolate 5 minute power off the limb (type I spicules that typically move up and down during their lifetimes) and a much more dynamic component (Type II spicules – often appear or disappear within a few seconds). Credit: De Pontieu *et al.* (2007b).

from the lower layers of the solar atmosphere to the corona and the solar wind. It is thus now generally accepted that mottles are jet-like structures and that they cover, at any time, about 1% of the solar surface.

The classical spicules have widths in the range 500–2000 km, reach heights in the range 5000–10 000 km above the photosphere through propagation typical speeds of ~ 20–50 km s^{-1}. Hence, they have lifetimes of just 5–10 minutes. On average, they are not vertical but at an angle of $\sim 20°$. Spicules are usually identified through Hα and Ca II lines, which yield their thermodynamic properties. Their temperatures are believed to be of the order of 7000–15 000 K, their electron densities of the order of $4 \times 10^{10} - 10^{11}$ cm^{-3} and their gas pressures of the order of 0.2 dyn cm^{-2}. Spin about the axes has been recorded (Beckers, 1968), although this has not received detailed attention.

A faster type of spicule has now been identified by Hinode Solar Optical Telescope observations (De Pontieu *et al.*, 2007b). The above slower type or *Type I* spicules are probably driven by shock waves that form when the energy from global solar oscillations and convective flows leaks into the atmosphere. The faster *Type II* spicules form extremely quickly, in about ten seconds, possess short lifetimes of 10–150 s, and propagate with speeds of order 50–150 km s^{-1}. They are consequently rapidly heated to at least transition region temperatures. These properties indicate that Type II spicules are produced by magnetic reconnection in the vicinity of magnetic flux concentrations.

The flow in spicules is magnetohydrodynamic. They all channel Alfvén waves with significant amplitudes of order 20 km s^{-1}. It is also well known that the arrangement of mottles is governed by the configuration of the local magnetic field. Thus mottles are usually organised into small groups, called chains, and larger groups, called rosettes. In a chain the mottles point in the same direction. Rosettes have a circular shape and owe their morphology to the presence of magnetic canopy-like flux tubes filled by the plasma ejected from below and streaming outwards from a common bright centre. Strong magnetic flux of the order of 10^{20} Mx and strength 1000 G have been detected in the bright centre of a rosette (Tziotziou *et al.*, 2003). The Hα width of a spicule slowly decreases with height. After reaching its final altitude, a spicule may either fade away or fall back along its initial path, although some spicules are observed to fall along a different arched path.

2. Dynamic fibrils (De Pontieu *et al.* 2007a). Dynamic jet-like features can also be seen in and around magnetically active plage regions. These features are observed in Hα and in Lyα as Lyman-alpha jets (Koza *et al.*, 2009). They are shorter (1400 km) and shorter lived (3.6 minutes) and appear to form a subset of what have traditionally been called active region fibrils. There are also fibrils that do not show jet-like behaviour. They are thought to be related to magneto-acoustic shock waves driven by convective flows in the photosphere. Dynamically, maximum speeds of 10–30 km s^{-1} are found, with decelerations of 50–250 m s^{-2} and other properties very similar to spicules. The deceleration correlates with the maximum speed and the duration correlates with the length.

3. Surges. Extreme ultraviolet plasma jets are associated with chromospheric Hα surges which typically take the form of a straight or slightly curved spike. They display twisting structures, often originating near one end of a pair of small flaring loops and appearing to open to space. The Hα surges, however, are smaller and only trace the edges of the jets. They occur later than the jets but have dark extreme ultraviolet counterparts appearing in the bright jets (Jiang *et al.*, 2007). The surges last up to 30 minutes, recurring with a period of one hour or more. They reach heights of up to 200 000 km with peak velocities of

50–200 km s^{-1}, and then fade or fall back to the chromosphere, apparently along the same trajectory as the ascent. The observed surges spin in a sense consistent with the relaxation of the twist stored in the magnetic fields of the moving magnetic bipoles (Canfield *et al.*, 1996).

There are clear similarities between spicules and surges. Both tend to be recurrent at the same position and both carry the signature of the underlying magnetic field. It is thus plausible that they are the manifestation of the same phenomenon occurring on different scales, the phenomenon probably being magnetic reconnection.

4. Giant chromospheric Ca [II] H jets (Nishizuka *et al.*, 2008). In February 2007, Hinode discovered a giant jet with two remarkable adjacent components at the solar limb. Simultaneous observations with the Hinode and TRACE satellites revealed that hot ($\sim 5 \times 10^6$ K) and cool ($\sim 10^4$ K) jets were located side by side and that the hot jet preceded the associated cool jet by 1–2 minutes. A structure like a current-sheet was observed in the optical (Ca II H), extreme ultraviolet (195 Å) and in soft X-ray emission, suggesting that magnetic reconnection is occurring in the transition region or upper chromosphere. Alfvén waves were also observed with Hinode. These propagated along the jet at speeds of ~ 200 km s^{-1} with amplitudes (transverse velocity) of ~ 5–15 km s^{-1} and a period of ~ 200 seconds.

5. Coronal EUV microjets (Gurman *et al.*, 1998) are extreme ultraviolet chromospheric jets that are identified with spicules but with lifetimes about ten times shorter.

6. Coronal jets. Coronal jets were originally found in association with macrospicules (Yamauchi *et al.*, 2004), eruptive events that are generally of columnar form. They can be considered to be giant spicules or small surges which reach heights of 7000–40 000 km above the limb with rise velocities of 10–150 km s^{-1}, and lifetimes of minutes. They were found to be transitory X-ray features with collimated motions associated with a range of other activity including sub-flares, micro-flares, X-ray bright-point brightenings and emerging flux regions (Shibata *et al.*, 1992). This suggests that jets and flares have a common physical origin. About half of the macrospicules are distinctly not erupting loops, but instead are surge-like jets showing a characteristic inverted-Y spike shape similar to the Eiffel tower.

The Hinode satellite has recorded a large number of coronal jets in X-rays, of order ten per hour (Cirtain *et al.*, 2007), and our knowledge has now substantially increased. The length of the X-ray jets is a few times 10^4 to 4×10^5 km and the speeds range from 10–1000 km s^{-1}. There appears to be two peaks in the velocity distribution: one near the Alfvén speed (800 km s^{-1}) and another near the sound speed (200 km s^{-1}). This suggests two distinct formation mechanisms, with the fast jets associated with magnetic reconnections, and the slow jets driven by magnetosonic shock waves. The jets extract mass of 10^{12}–10^{14} g and thermal plasma at a temperature of a few million kelvin, giving a thermal power of the jets of 10^{27}–10^{29} erg s^{-1}. This power has to be released on a relatively short timescale, of the order of several minutes, much smaller than that for energy injection in the corona. Thus, the major source of the energy is again likely to be magnetic. Bain and Fletcher (2009) also reported strong evidence for accelerated non-thermal electrons, consistent with jet models based on magnetic reconnection. The first observations of a jet in hard X-rays (Bain & Fletcher, 2009) also revealed the presence of thermal plasma at a temperature of about 28×10^6 K.

These jets occur in coronal holes and it has now been confirmed that the coronal jet plasma is indeed ejected along open field lines (He *et al.*, 2010). The ultimate source of a jet has been traced down to supergranular converging motions. The motions drive a shock wave which drives upwards and pushes the jet plasma with it. The magnetic evolution has been studied at the jet base in more detail. It was found that the jet launching arises either when two flux tubes of opposite magnetic polarity interact or when several flux tubes with identical polarity are squeezed (He *et al.*, 2010).

7. *White-light polar jets* (Wang & Sheeley, 2002) White-light ejections are the outward extensions of extreme ultraviolet jets, which originate from flaring bright points. The white-light jets have angular widths of $\sim 5°$ and speeds of order 600 km s^{-1}. Many of the ejections are recurrent in nature and originate from active regions located inside or near the boundaries of non-polar coronal holes as well as polar regions. The jets are launched from regions which contain systems of closed magnetic loops partially surrounded by open fields. Perturbations in the closed fields caused them to reconnect with the overlying open flux, releasing the trapped energy in the form of jet-like ejections. In some events, the core of the active region erupts, producing fast, collimated ejections with broader openings.

Rapid progress in the field of solar jets is being made with the Solar Dynamics Observatory as well as Hinode, both space missions. They have helped reveal the importance of spicular jets to the energy budget of the corona and the solar wind. Up to one hundred times more mass is circulated through the jets into the upper chromosphere than actually exits the Sun in the wind (De Pontieu *et al.*, 2011). Most of the plasma is heated to temperatures approaching 100 000 K, while a small fraction reaches temperatures above 1 MK, making a contribution to coronal heating that could solve long-standing problems.

8.5   Summary

The significance of jets within the solar system is only just being realised. We are learning that both hydrodynamic and magnetohydrodynamic mechanisms have their place. The jets are ephemeral, often fountain-like, returning some material to the surface. Some of them are recurrent.

Jets emanate from the surface of comets as they approach the Sun. Several space missions in the last decade have provided the close encounters necessary to locate and resolve the jets as they leave the surface on the scale of hundreds of metres. Further missions may help distinguish sub-jets and the properties of the launching landscapes. Clearly, however, solar heating can trigger geyser-like gas jets without the need for either a magnetic field or an accretion disc.

Cometary jets are indeed puzzling. How can well-collimated jets of gigawatt power be produced from an irregularly shaped rock with little gravity or atmosphere? What holds the jet direction stable? Sublimation of water ice clearly occurs during the Sun approach. However, the vapour should expand freely into the vacuum over a large solid angle.

Solar jets have long been studied in the form of spicules which transport sufficient mass to supply the solar wind one hundred times over (Sterling & Hollweg, 1988) but are thought to be too cool to power the corona. However, X-ray data have revealed accompanying hot plasma jets on diverse scales and speeds, generated by more than one type of reconnection mechanism. These hot jets are capable of resupplying the corona and, hence, may provide the means of balancing the energy budget in the corona (De Pontieu *et al.*, 2011).

9

Jet launching

Theorists have found the field of jets to be a lucrative playground. Chimneys, funnels, tunnels, vents and nozzles evoke common structures that we know can be responsible for launching collimated flows. However, to work on cosmic scales requires alternative conditions and unfamiliar physical regimes to be explored. Severe constraints limit the number of tenable models to just a few.

There are three challenges to face: to launch, to accelerate and to collimate. Winds can be launched and accelerated relatively easily with thermal, radiative and magnetic driving from stars and accretion discs. Vast amounts of energy are released through either collapse, external heating, mass infall or nuclear fusion. Upon expansion into the surroundings, a fraction of the released energy is in some combination of released gas, radiation and magnetic field. During transport, it is converted into radially directed kinetic energy. The discovery of jets, however, indicates that the theory is far more complex than just demonstrating the existence of a driver.

This chapter will first describe pure hydrodynamical methods for driving jets as originally invoked for the extragalactic case, but now relevant to cometary and planetary jets. The observations we have so far discussed have indicated that accretion discs and jets are very strongly correlated except in the solar system. In light of this disc–jet connection, models have concentrated on the need for a rotating accretion disc and associated magnetic torques. Differences come in ascribing the origin of the magnetic fields to the disc itself or to the central star. The most developed of these models are described below in detail. These are based on centrifugally driven disc winds and X-winds. However, there are many variations and alternatives which are then discussed towards the end of this chapter.

The spin paradigm uses a magnetic field as a means to tap the spin of a black hole. The jet power is extracted electromagnetically and so is expressed as a Poynting flux. This energy is subsequently dissipated on large scales. Hence, rather than the accretion rate, it is the black hole spin which may determine whether or not extragalactic radio jets are produced. This is an attractive basis for understanding the observed dichotomy between radio-loud and radio-quiet active galactic nucleus.

Testing the models requires predictions. As already reviewed, we have some data on the shape, the rotation and the acceleration, but only in regions remote from the initial launch. The result is that rigorous tests have not been forthcoming until now. There are too many unknown physical factors and intervening processes which disguise the original launch process.

9.1   Hydrodynamic methods

9.1.1   *Hydrodynamic methods: nozzles*

A purely hydrodynamic principle of jet formation is encapsulated in the de Laval nozzle mechanism. It has required substantial modification to find applications in astrophysics. It was first suggested in the context of radio galaxies even before it had been established that jets were ubiquitous (Blandford & Rees, 1974). It was later adopted as the basis for launching stellar jets and, closer to the original engineering concept, has now been applied to explain cometary and planetary jets.

The de Laval principle was developed and applied by Swedish inventor Gustaf de Laval in 1897 for use in some types of steam turbine. It is now used extensively in rocketry. In this form, a single supersonic jet is generated when compressible gas of high pressure is injected at one end of a nozzle, i.e. an engineered tube that is pinched in the middle. As the gas accelerates down the tube, it loses pressure. With the correct design, the flow becomes transonic exactly at the constriction, where the nozzle is narrowest. As the nozzle expands, the falling pressure is accompanied by further acceleration.

The nozzle converts heat energy into directed kinetic energy. Its operation relies on the contrasting behaviour of gases flowing at subsonic and supersonic speeds. It is best demonstrated by taking a steady one-dimensional flow of an isentropic gas with pressure p, density ρ and specific heat ratio γ. The flow of speed u is confined to a narrow channel of cross-sectional area A. Therefore, the mass flow rate down the channel, $\rho A u$, is constant. The momentum flow rate is also constant and consists of two terms related to the ram pressure and the thermal pressure. In terms of the momentum flux, the force along the direction z is $\rho u(du/dz)$. We equate this to the pressure gradient:

$$\rho u \frac{du}{dz} = -\frac{dp}{dz}. \tag{9.1}$$

With an adiabatic equation of state, the energy equation is $p = \rho^{\gamma}$ and the sound speed is given by $c_s^2 = \gamma p/\rho$. We may therefore eliminate the pressure to yield

$$\rho u \frac{du}{dz} = -c_s^2 \frac{d\rho}{dz}. \tag{9.2}$$

Finally, the constant mass flow rate is written in the form

$$\frac{d\rho}{\rho} + \frac{du}{u} + \frac{dA}{A} = 0 \tag{9.3}$$

and used to eliminate $d\rho$. The neat final result expresses the speed purely as a function of area A:

$$(1 - M^2) A \, du = -u \, dA, \tag{9.4}$$

with the Mach number $M = u/c_s$. Hence, the speed of a subsonic flow of gas will increase if the channel carrying it narrows because the mass flow rate is constant. At subsonic flow speeds, the gas should be thought of as compressible; sound waves, or small pressure waves, will propagate rapidly through it.

At the 'throat', where the cross-sectional area is a minimum, the gas speed locally is sonic (Mach number $= 1.0$), a condition called choked flow. As the nozzle cross-sectional area increases, the gas begins to expand and the gas flow increases to supersonic velocities. In this downstream section, sound waves cannot propagate backwards through the gas as viewed in the frame of reference of the nozzle (Mach number > 1.0).

This smooth adiabatic transition is only one possible flow solution. The ratio of gas pressure to the ambient pressure at the exit is crucial to the flow type. A de Laval nozzle will choke at the throat only if the pressure and mass flow through the nozzle are sufficient to reach sonic speeds, otherwise no supersonic flow is achieved and it will act as a Venturi tube (a tube which constricts a flow). In addition, the pressure of the gas at the exit of the expansion portion of the exhaust of a nozzle must not be too low. Because pressure waves cannot travel upstream from the exit through the supersonic flow, the exit pressure can be significantly below the ambient pressure it exhausts into, but if it is too far below ambient, then the flow will cease to be supersonic, or the flow will separate within the expansion portion of the nozzle, forming an unstable jet that may 'flop' around within the nozzle.

Solid walls of a specified shape are not necessary for supersonic jets to be generated. Instead, a jet can be produced when compressible gas of high pressure is injected into a dense cloud of gas. In this case, the gas expands as a spherical wind which accumulates within a cavity of growing size. The cloud pressure will eventually halt the expansion and the walls can be gradually sculpted into the optimum bubble shape by the interacting media. The optimum shape is one where the two gases represent inviscid fluids separated by a thin viscous interface which approximates to a tangential discontinuity. The interface of the cavity with the cloud may distort and channels develop through which the hot gas accelerates and escapes. The imposed conditions on the injected and confining gases for jet launching to occur, as envisaged in the twin-exhaust model of Blandford and Rees (1974), were based on the de Laval mechanism. However, in the classical form, this mechanism generates a single jet. In the twin-exhaust, a small axisymmetry in the cloud suffices to bifurcate the flow into two antiparallel axial jets. The axisymmetry can be caused by rotation or gravitational effects. An alternative symmetry could result in a thin, supersonic, sheet-like jet of solar-wind material entering far into the oncoming interstellar wind at the heliopause (Opher *et al.*, 2004).

In the Blandford and Rees (1974) model, hot gas of high pressure is continuously injected into a compact central region of a uniform cloud. The gas expands radially as a spherical wind, converting the thermal energy into supersonic wind energy. However, the radial expansion reduces the density and, hence, the wind ram pressure until equilibrium is reached with the ambient pressure (Fig. 9.1). This defines a zone in which the wind energy is converted back into thermal energy through a standing spherical shock wave. The hot shocked gas then forms the cavity between the wind shock and the ambient medium. Because it is hot, however, an axial pressure gradient will efficiently push the cavity material towards the axial direction.

A major issue with the model is the stability of the cavity and the throats. The Rayleigh–Taylor instability occurs if the confining cloud is supported either by self-gravity or by gravity (of stars or a black hole), inducing the infall of dense filaments into the cavity (Norman *et al.*, 1981). The instability tends to remove large chunks of hot gas from the walls, which then disrupt the smooth cavity flow. The Kelvin–Helmholtz instability is effective on the throats since this operates across shear layers (see Section 10.5). This is potentially completely disruptive since the non-linear growth of long-wavelength modes pinches the jet. To suppress this, the gas flow out of the cavity must beat the Kelvin–Helmholtz pinching.

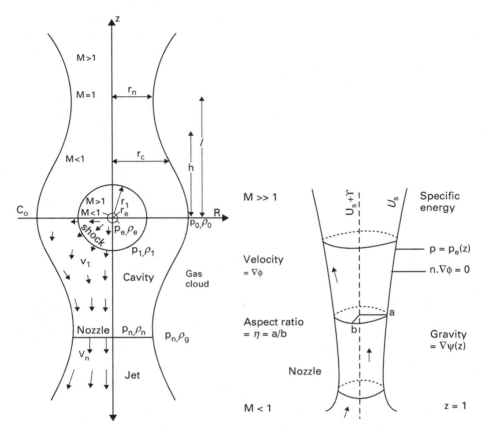

Fig. 9.1 Key features of the classical de Laval nozzle arrangement as modified to apply to astrophysical jets in which energy is continuously supplied from a central engine into an almost uniform environment with symmetry about the $Z = 0$ midplane. Cylindrical coordinates are used. Extracted from Smith *et al.* (1983) and Smith (1994b).

Blandford and Rees (1974) proposed that the model would also work for a relativistic plasma, with a flow speed at the throat of $c/\sqrt{3}$ (the sound speed of an ultra-relativistic plasma) in the ultra-relativistic case. The fluid approach is immediately justified if a magnetic field, B, is present since the Larmor radius of a proton is just $\sim 10^{-12}\gamma/B$ pc where the field is measured in Gauss and γ is the proton Lorentz factor. Hence, in the extragalactic case, the field just has to be present but does not have to be dynamically significant.

The first numerical simulations of the twin-exhaust identified three regimes: (i) bubble release from the throat, (ii) smooth jet outflow and (iii) large clump-cavity oscillations (Norman *et al.*, 1981). These cylindrically symmetric calculations showed that the critical parameter relates the jet power to the internal energy of the confining cloud. The jets remain globally stable over a quite narrow range of about 40 for a fixed confining gas pressure. As a result of the strong observational constraints (Smith *et al.*, 1981; Wiita & Siah, 1986), the application to extragalactic jets has not been strongly pursued, although the mechanism itself may operate over a quite extended parameter range (Frank & Mellema, 1996).

Hydrodynamic launching has also been considered in the context of young stars. Rather than the late steady-state phase, Königl (1982) considered the expansion phase in which a turned-on spherical wind is blowing an expanding bubble and sweeping up the ambient medium into a dense, cold accelerating shell. An axially flattened mass distribution with a sharply falling density profile may then generate twin jets. If the density within the bubble is sufficiently high, then the shocked wind cools before escaping. In this case, the shell is momentum-driven by the wind ram pressure in a snowplough phase. Eventually, however, the wind density falls through expansion and the bubble becomes energy-conserving. In this phase, the internal energy is available to be converted into directed jet energy provided a de Laval mechanism can operate. This model version has the advantage that the slowly advancing shell provides the confining wall for the rapid alignment of the fast wind.

It should be noted here that the mass distribution must be highly flattened. While the confined bubble is still expanding, there must be considerably less resistance along the axis for the jets to drive out material and emerge on a much larger scale. This is possible, but leads to flows best described as champagne flows or wide-angle winds rather than jets. In this context, an axial magnetic field can instead provide the ambient pressure to collimate a wind, but is still only effective once the wind-blown bubble has stagnated (Matt *et al.*, 2003).

Hydrodynamic simulations of the interaction of a central wind with a highly flattened environment, characterised by a toroidal density distribution, were performed by Frank & Mellema (1996). The simulations revealed time-dependent non-linear features including a prolate wind shock and a chimney of cold swept-up ambient material dragged into the bubble cavity. The collimation into a supersonic jet combines both de Laval nozzles and focusing of the wind via the prolate wind shock.

A hybrid jet model for young stars was proposed by Smith (1986). If the outflowing wind is dense, it may well be radiative rather than adiabatic. In this case, it had earlier been shown that the injection of hot gas into an axisymmetric slightly flattened cloud would generate a wide barrel or ovoid-shaped outflow which may become focused towards the axis (Canto & Rodriguez, 1980). The inner spherical wind is deflected towards the axis through an oblique reverse shock. The shocked wind cools into a thin supersonic shell which forms a dense moving interface between the shock and the ambient cloud. The exact configuration, sketched in Fig. 9.2, is thus found by balancing at each point the ambient pressure with the sum of the ram pressure component of the deflected wind and the centrifugal force of the curved supersonic shell. The result is a focusing of the outflow, leading to a supersonic impact on the axis. Hence, it is a natural means to produce those Herbig–Haro objects which appear fragmented rather than bow-shaped and may even create quite well-collimated jets (Canto *et al.*, 1988).

The candle-flame model adapted the above scenario by assuming an inner torus or accretion disc which anchors the shock wave in the symmetry plane (Smith, 1986). The collimated flow geometry then stems from two conditions: firstly, from an oblique interaction at the disc surface and, secondly, from a quite flat fall-off of density with distance. The result is a more gradually focused flow and high collimation into a narrow flame-like jet. The oblique deflection implies that a large fraction of the wind momentum can be channelled into the jets. However, it is not clear how jet-like such a flow would be with the thin shell subject to strong shear instabilities.

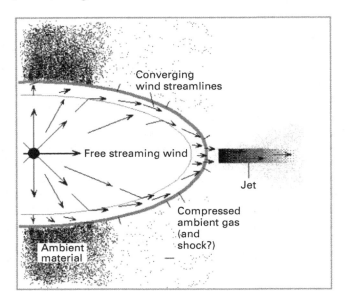

Fig. 9.2 Jet formation via a conical converging flow as sketched by Frank (1999). The ambient medium confines a wind into a prolate momentum-conserving bubble. The shocked wind forms a thin shell of material that streams along the interface, converging towards the axis. A second shock deflects the material into a jet.

Dense outflows may indeed be radiative rather than adiabatic, especially early in the evolution of a stellar wind. In this case, an expanding wind-blown cavity is inertially confined rather than thermally confined. Furthermore, numerical simulations indicate that jet-type flows may arise from such a mechanism (Mellema & Frank, 1997). Nevertheless, hydrodynamic models have not been pursued recently for young stars for several reasons. One reason is the high jet speeds measured, which require models that predict launch locations very close to the central object. Secondly, the stability has not been sufficiently questioned, despite some numerical simulations. Thirdly, the formation of the massive central wind is not explained within the model.

9.1.2   *Hydrodynamic methods: discs*
A model in which the launch takes place as material from the inner edge of an accretion disc impacts the central object was proposed by Torbett (1984). A boundary layer shock is set up in the disc which creates a high thermal pressure zone embracing the star. A small fraction of the accreting mass escapes predominantly in a direction perpendicular to the disc plane. Since a central surface is required, the model does not apply to accretion onto black holes (see also Stute *et al.*, 2005b). Nearly half of the total available accretion energy can be released in the boundary layer if the central object is slowly rotating. This energy is released in a strong shear layer which is set up between the disc and surface. The shear inevitably winds up a toroidal magnetic field, which would help to collimate the outflow (Pringle, 1989) but not necessarily aid in the acceleration (Soker & Regev, 2003).

If the heated gas remains hot, a corona forms and a wind may then be thermally driven. On the other hand, if cooling is effective, the boundary layer will be narrow and radiation from the layer may be transferred into both the adjacent disc and outer stellar surface layer. For reasonable densities in a protostellar disc, Torbett & Gilden (1992) concluded that the mechanism fails to accelerate matter to the escape velocity due to radiative cooling. Nevertheless, a small fraction of the accreted matter, of order 1%, can still be accelerated to a significant fraction of the orbital speed.

Soker & Lasota (2004) apply this mechanism to systems containing disc-accreting white dwarfs to explain the absence of jets in cataclysmic variable stars. For the thermal model to operate, the accreted material must be strongly shocked as a result of large gradients of physical quantities in the boundary layer. It may then only cool on a timescale longer than its ejection time from the disc. Ejection of mass requires high mass-accretion rates, exceeding 10^{-7} $M_\odot$ yr^{-1} and 10^{-6} $M_\odot$ yr^{-1} for young stellar objects and white dwarfs, respectively. Such high accretion rates are not associated with cataclysmic variables. In any case, however, large hot bubbles rather than jets may be ejected. In the SPLASH scenario (SPatiotemporal Localized Accretion SHocks), a succession of locally heated bubbles expand from the boundary layer and merge (Soker & Regev, 2003). This plasma may accelerate to speeds in excess of the escape speed. As might be expected, support for hot bubble formation has been forthcoming from numerical simulations, provided the accreting material has nowhere else to go (Stute *et al.*, 2005). Collimation has yet to be achieved.

9.1.3 Hydrodynamic methods: vents

Comets turn into fountains of gas and dust as they approach the Sun. Solar energy vaporises the outer layers of the nucleus and heat is conducted into caverns of pressurised gas. The gas escapes through fissures and erupts into jets. Such a hydrodynamic scenario based on regulation by thermal pressure is undoubtedly responsible for cometary and planetary jets.

The collimated-jet model considers isolated active regions on the nucleus that eject gas into the coma to produce diverse spatial features (Farnham, 2009). Each jet is expected to be unique and so provides a natural explanation for the different properties (density, outflow velocity, etc.) that are often distinguishable in the structure within the coma. Surface regions of high gas production can be attributed to inhomogeneities in the composition of the sublimating ices which drive the activity, or they may represent the nature of the vents in an otherwise impermeable mantle. The combination of radial ejection and comet rotation produces linear features on small scales but curves, spirals and helical structures on large scales according to the viewing angle and the latitude of the active region.

The presence of collimated jets requires that the gas emission must be insignificant over large portions of the nucleus. An inactive mantle in the form of a regolith-like layer that may be quite thin with low tensile strength is expected. This mantle is dark, and a low thermal conductivity effectively insulates this thin regolith, best described as a black crust. Thus, most of the comet's surface is insulated and the volatile material underneath is protected from sublimation.

On the other hand, we expect low-speed grains emitted from active regions to migrate to other spots and settle back onto the surface (Kitamura, 1986). Thus, a jet in one spot may be strong enough to keep from choking itself off, but the ejected material can contribute to mantle formation elsewhere. As discussed in Section 7.8, bright patches representing water deposits must be recent depositions that are being evaporated. That is, water ice patches must be transient, and are probably recondensation from residual jet emission that continues at lower speeds during the night.

Alternative mechanisms have been proposed but remain speculative, not having been sufficiently developed. The idea that precursor ions of HCN would be accelerated by energetic electrons in ionospheric current sheets or current arcs (Ip & Mendis, 1977) can be generally excluded after the recent imaging of cometary nuclei.

The sublimation of ices directly from the surface layers or after percolation through a surface layer, especially from within shaped pockets, has also long been considered (Keller *et al.*, 1994). The problem is that the production of highly collimated jets, rather than wide jets and fans, requires a contrived configuration and has no quantitative basis. This is because random thermal motions will cause a stream of gas to rapidly diffuse laterally when it issues into a diffuse environment. In a collimated flow, the radial velocity of the particles must be much larger than the random thermal motions to allow the stream to retain some degree of collimation. When the gas has expanded so that it is not collisional, it enters a pressure-free particle regime. Hence, the gas flow is susceptible to the random thermal motions near the comet's surface where the density and temperature are both high. The dust flow, on the other hand, can decouple from the gas at a few nuclear radii, making it much easier to retain collimation. However, even if it has decoupled from a perfectly collimated gas flow, the dust particles retain speeds of order of the gas thermal speed (Yelle *et al.*, 2004).

One of the earliest mechanisms proposed for launching collimated jets was based on the deeper topography of the surface as sketched by Whipple (1989). Trenches and deep craters can produce a 'lens' focusing effect, where the insolation from the Sun is reflected and focused by the opposing surfaces (Colwell *et al.*, 1990; Keller *et al.*, 1994).

The floor and sides of a crater heat up, increasing the sublimation rate (Colwell *et al.*, 1990). The interaction of the gas flowing from these offsetting surfaces also acts to focus the outflowing gas, producing a collimated stream. The lens effect may produce positive feedback by increasing the sublimation and so may cause the crater to deepen, intensifying the effect even further. However, at some point self-shadowing will limit the effect.

It has been suggested that cometary dust jets, as for example imaged by the Halley Multicolour Camera (Fig. 8.2), may originate from such crater-like surface features which surround inactive central (core) regions (Keller *et al.*, 1994). The decrease of pressure above the non-sublimating core surface tends to converge the surrounding gas flow. Dust accelerated in streams from such configurations may possess enhanced collimation. This would concentrate large particles into radial filaments. Dust liberated from the bottom of the crater emerges at the surface level with a finite velocity. These particles have a larger outward (radial) momentum than particles leaving the surrounding surface with an initial velocity close to zero. However, each particle also retains a high lateral component from the same gas-grain acceleration process and well-collimated jets are not plausible.

The mechanism usually invoked for collimating flows is related to the de Laval nozzle and the process is termed the Venturi effect. The focusing is especially pronounced for the dust, which is more likely to remain collimated, even if the gas expands laterally after exiting the nozzle at a low Mach number. The Venturi effect may also be produced in 'capillary tubes' in the mantle, with each tube acting as a miniature nozzle (Skorov & Rickman, 1995).

If a number of these tubes are distributed around a region on the surface (e.g. an active area) then the superposition of the individual tubes will contribute to what is observed as a single jet.

Furthermore, the lateral diffusion of the flows from around the perimeter of the active area has a component that converges inwards. This inward flow generates a ram pressure that will reduce the lateral diffusion of the sources near the centre and increase the degree of collimation of the overall jet. Using laboratory experiments, Thiel *et al.* (1991) showed that gas percolating through a matrix surface does seem to induce collimation in the outflowing dust, which supports the possibility that this may be one means of producing jets.

The explosive escape of volatile materials soon became the accepted model, supported by Whipple and Houpis (Whipple, 1989). The material is initially cool and trapped in pockets or chambers below the surface. The crust helps the comet absorb heat which, as the Sun is approached, causes some of the trapped ice under the crust to sublimate. With pressure building beneath the crust, the weakest areas of the crust shatter, and the gas shoots outward like a geyser. Any dust that had been mixed in with the gas is ejected by the gas drag. From this moment, solid confining walls channel the gas flow and the wall shapes can be gradually sculptured into a streamlined form by the abrasive effect of the jets. Finally, the entrained dust particles scatter sunlight to produce the observed jets.

The geyser-nozzle model as shown in Fig. 9.3 is the most developed interpretation of cometary jets. Numerous such pockets of gas are created from the heated ice in the caverns. Cracks appear as the gas is forced by the excess pressure to pass through to the surface, reaching supersonic speeds after the narrowest point (Yelle *et al.*, 2004).

Once a steady state has been reached, the jet mass outflow rate will equate to the rate of ice sublimation. This mass sublimation rate will depend on four parameters: the heated

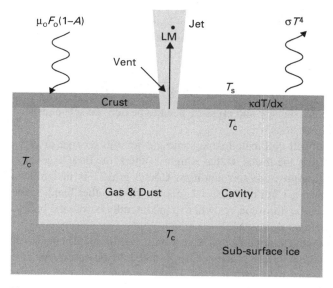

Fig. 9.3 The formation of cometary jets through geysers supported by sub-surface high-pressure gas. The gas is produced as ice below the crust is sublimated (Yelle *et al.*, 2004).

surface area of the crust lying over the cavern, A_c, the incident solar flux, F, the latent heat of sublimation of ice, L, and the fraction of the incident flux, ζ, which is conducted into the comet rather than reflected or re-radiated. In terms of reference values, we obtain for the sublimated mass

$$\dot{M}_{subl} = 20 \left(\frac{\zeta}{0.5} \right) \left(\frac{F}{10^6 \, \text{erg s}^{-1} \, \text{cm}^{-2}} \right) \left(\frac{L_{subl}}{2.8 \times 10^{10} \, \text{erg g}^{-1}} \right)^{-1}$$
$$\times \left(\frac{A_c}{10^6 \, \text{cm}^2} \right) \, \text{g s}^{-1}. \tag{9.5}$$

Consequently, if the ice has a density of 0.5 g cm^{-3}, a depth of one metre of ice will be sublimated in about one month.

The density of the resulting gas depends on the the vapour pressure, which depends critically on the rate of escape of the gas as well as on the temperature on the roof of the cavern. The latter is rather uncertain, depending on the conductivity and regolith depth, but plausible estimates can be made. The calculations of Yelle *et al.* (2004) yield the following orders of magnitude for a comet approaching the inner solar system: a cavern temperature of 200 K, density 10^{-6} g cm^{-3} and pressure 1000 dyn cm^{-2} (corresponding to 100 Pa). In comparison, the bulk modulus of the powdery-icy mixture which makes up the comet may be 1000 Pa, providing some resistance to sustain the reservoir. It is not absolutely clear, however, that this would be sufficient to support the underlying gas for a lengthy period.

The flow pattern as the gas passes through the nozzle is exactly as in Eq. (9.4) provided a one-dimensional flow can be presumed. For an isothermal flow, the isothermal sound speed must of course be taken. The ambient pressure will be considerably lower, ensuring that supersonic speeds will be achieved beyond the throat of the nozzle.

The actual dimensions are uncertain. A single isolated jet may have a throat diameter of just 10 cm within a vent of depth 1 m through which material from a cavern of diameter 10 m may be ejected. Given the sound speed of the H$_2$O gas of $\sim$0.5 km s^{-1}, we immediately obtain a rough estimate of the rate of escape of gas that is consistent with that given by the ice sublimation:

$$\dot{M}_{vent} = 20 \left(\frac{\rho}{10^{-6} \, \text{g cm}^{-3}} \right) \left(\frac{u_{vent}}{5 \times 10^4 \, \text{cm s}^{-1}} \right) \times \left(\frac{A_{vent}}{200 \, \text{cm}^2} \right) \, \text{g s}^{-1}. \tag{9.6}$$

In the general case, the gas density and pressure need to be derived self-consistently given the restricting throat dimensions.

Here, we note that a well-collimated supersonic gas jet will emerge provided the gas remains confined well after the throat so that a highly supersonic flow is generated before the gas exits into the low-pressure surroundings. Comet gravity is insignificant, with a typical escape speed from a 1 km nucleus of 1 m s^{-1}. On the other hand, if the throat is very close to the surface, the transonic gas will expand laterally before the ballistic motion fully dominates the thermal expansion.

To date, it is the associated jets of dust that have usually been detected close to the nucleus and, hence, it is not clear that the gas is also well collimated. In the model, the dust particles are accelerated by collisions with gas particles. The mean free path of molecules is much less than the vent dimensions, validating the fluid approximation. However, the mean free path is much larger than the typical dust particle size. This implies that the gas flow is free molecular with regards to the dust. Small dust particles may be swept along with the gas, possibly ending up in a wider fan if the gas expands after ejection. Larger particles will be accelerated by the gas through the throat to a lower speed, and may then decouple from the gas due to the consequent gas expansion and lowered collision rate. Since many individual gas–grain collisions are necessary to accelerate the heavy grains, the grains maintain their highly directed motion after decoupling.

The opening angle of the dust jet will depend on the particle size. The speed of the dust jet depends on the gas parameters with micrometre-sized particles expected to achieve speeds of ~ 50 m s^{-1} through the vent. However, within an expanding jet or wind, acceleration due to drag continues immediately above the surface. With spherical expansion, the drag force would not be effective in increasing the particle speed. Dust speeds exceeding 500 m s^{-1} require either strong outbursts (Gombosi & Horanyi, 1986) or the presence of a collection of many vents covering a surface area of, perhaps, a square kilometre. Grains caught within this larger venting field are likely to experience enhanced acceleration.

The grain mass outflow can be lower than the gas mass outflow (e.g. a ratio of 0.27, Gombosi & Horanyi (1986)) or higher (e.g. 2.3, Veal *et al.* (2000)). If we take a total mass outflow rate of 2×10^7 g s^{-1} over the surface of a comet with a radius of 3 km, then venting through a total area of 10^{-4} of the surface would be required. Rather than the small vents considered above, images of the Comet Tempel nucleus (Fig. 8.7) indicate that we should in fact scale up the jet system and take instead 10 m vents supplied from sub-surface ice covering a square kilometre.

Although there is currently no accepted mechanism for jet production from comets, further work on the geyser model presents opportunities to predict jet behaviour in the coming years.

The total power of the Enceladus jets is considerably less than in the powerful cometary outflows. The mass outflows of 2×10^5 g s^{-1} and jet speeds of 0.3 km s^{-1} imply a power of under 50 MW, as compared with 10 GW for active comets. The basic venting mechanism described above also applies to these jets, providing an explanation for the supersonic flow. The vents in the ice are described as fissures or sulci of maximum size of 10 m, which can be shaped by mineral deposits into perfect de Laval nozzles. Here, however, the power source cannot be solar, but is consistent with deep interior heating as a consequence of planetary tidal shear as shown in Fig. 9.4. The reservoir may now be liquid water rather than ice due to the internal source of heat, but this is not clear (Kieffer *et al.*, 2009; Waite *et al.*, 2009).

If an initial state exists in which there are no cracks available to discharge the gas, the heating and ice sublimation would gradually raise the gas pressure, which would directly exert a pressure on the solid ice above. The bulk modulus of the strong ice structures on the moon is estimated as 10^6 Pa, which would suggest that gas densities may have to build up to 0.01 g cm^{-3} before forcing a way through. With this release, the gas pressure then falls back and a steady state can be attained where sublimation through internal heating balances gas escape. To quantify the steady state, we could assume that there are $N = 10$ vents of size 100 m^2 on the south polar terrain:

$$\dot{M}_{vent} = 3 \times 10^5 \left(\frac{N}{10}\right)\left(\frac{\rho}{10^{-6} \text{g cm}^{-3}}\right)\left(\frac{u_{vent}}{0.3 \times 10^5 \text{ cm s}^{-1}}\right)^{-1}$$
$$\times \left(\frac{A_{vent}}{10^6 \text{ cm}^{-2}}\right) \text{ g s}^{-1}. \tag{9.7}$$

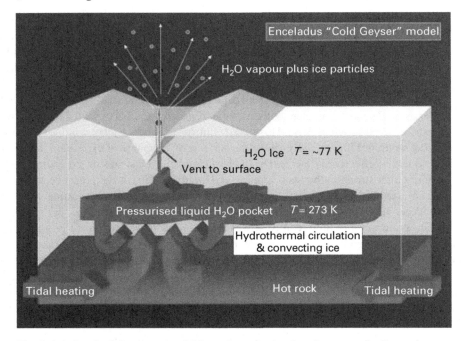

Fig. 9.4 A sketch of the scenario which produces ice jets in solar system bodies such as Enceladus. Credit: NASA/JPL/Space Science Institute.

In this case, the normal pressure of the sub-surface gas is very low in comparison with the tensile strength of the surface ice.

Three models involving cryovolcanic processes should be noted. The Cold Faithful model is based on an explosive boiling of cold liquid water a few metres below the surface; the underlying pressure exerted on the ice crust is suddenly released due to an opening crack (Porco *et al.*, 2006). In this model, the maximum speed of ice grains, originating from water splashing in explosive boiling, is far too small to explain the observed plume properties. Moreover, observed nitrogen and methane cannot be in the liquid water at the low pressures involved. In the Frigid Faithful model the existence of a deep shell, of clathrates below the south pole of Enceladus is conjectured (Kieffer *et al.*, 2006). Clathrates can decompose explosively when exposed to vacuum through a fracture in the outer icy shell, releasing latent heat from dissociated clathrate. Nitrogen, methane and carbon dioxide gas may then be released along with ice particles. This model also runs into gas–grain momentum transfer problems. Finally, in the Frothy Faithful model (Fortes, 2007), dissociation occurs in warm rising cryomagma. This can account for the high ejection speeds but does not explain the magma that is needed to generate it. All these models concentrate on the physical heating process rather than the dynamics of collimated jet launching.

9.2   Jets via magnetic reconnection

Activity on the surface of a convective star is dominated by magnetic waves and magnetic reconnection events. A complex changing network of open and closed field lines

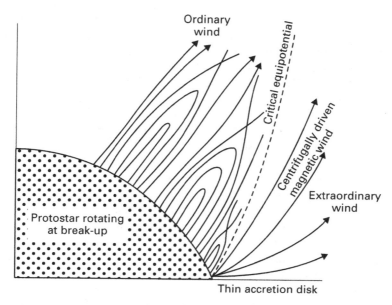

Fig. 9.5 Schematic diagram of the O-wind model as presented by Shu *et al.* (1994). The wind blows out along open field lines. Reproduced by permission of the AAS.

generates a stellar wind. In the case of protostars and young stars, this may produce an intense ordinary wind, or O-wind (see Fig. 9.5). In solar-type stars, a strong stellar wind is driven. It may well be that the wind is the cumulative result of the acceleration and heating of a large number of individual jets on all scales, from spicules and mottles to coronal jets. For convenience, we split the discussion here into these two extremes.

9.2.1 Spicules

Many theoretical ideas for generating solar spicules and mottles have been proposed. These mechanisms were not easy to develop or exclude because the spicule speeds are of order of the sound and Alfvén speed in the lower chromosphere, requiring non-linear methods or high computational power to test (Sterling, 2000). The basic models usually take a rigid tube of magnetic flux, which may expand with distance. Energy is pumped into the tube from unspecified events, which may cause a series of pulses of pressure or piston-like velocity variations. The major categories for models are based on fluid instability, on hydrodynamic waves, on rebound shocks, on Alfvén waves, and on magnetic field line reconnection. There are also many interesting models which remain in a separate speculative category. For example, they were considered to be self-channelled proton beams by Lorrain & Koutchmy (1996). The beam is excited by a dynamo situated either below or above the beam. In falling back, a cold surrounding sheath provides a return current of protons that cancels the outgoing current.

Shock waves have long been related to spicules but may find their home in the context of dynamic fibrils. Photospheric acoustic shocks alone cannot explain the high velocities of the surges in particular. Spicules were, however, modelled as ionising shock waves by Ashbourn and Woods (2005). The shocks originate in the chromospheric network and propagate upwards along flux tubes. The shock wave is driven by a circulating current resulting from electric forces within a flux tube and the ambient plasma. The current rapidly evolves to form a transverse current sheet which overtakes and heats plasma in a switch-on shock to generate $H\alpha$ emission. It may then reverse and retrace its path, further heating the plasma in the flux tube.

In the rebound shock model, the jet flow is collimated by a strong magnetic field in the form of a rigid flux tube (Sterling & Hollweg, 1988). Thus, torsional Alfvén waves are launched from the photosphere and steepen into slow MHD shocks. The shocks rebound off the transition layer and return due to gravity, compressing the atmosphere. The upward flow consists of a series of shock waves interacting with the previously compressed layers and driving the transition layer upwards. The general failing of shock models has been emphasised by related numerical simulations. These show that the low speeds of spicules inevitably produce ballistic motions in the solar gravitational field and so fail to account for their constant-velocity (non-ballistic) behaviour.

Reconnection of magnetic field lines is the most attractive candidate to explain spicules (Pikel'Ner, 1969), the explosive spicules in particular (Heggland *et al.*, 2009). The mechanism is now strongly favoured since it only requires oppositely directed magnetic field lines to be driven together. Early on, Uchida (1969) discussed how reconnection may release plasma bubbles that would be channelled through de Laval nozzles to generate supersonic jets. Simulations in one dimension generally assume a rigid flux tube geometry, and do not treat the magnetic reconnection explicitly but assume a sudden deposition of energy at a specified height. Two-dimensional simulations of explosive events have also been performed by Karpen *et al.* (1995) with a complicated initial magnetic field geometry. Shearing the field which contains an X-point leads to the development of magnetic islands, which can explain some intermittent behaviour of spicules.

Reconnection jets were given a 2.5-dimensional magnetohydrodynamic treatment by Takeuchi and Shibata (2001), motivated by strong magnetic flux cancellation at the photospheric level. In the photosphere, convection intensifies the weak magnetic field, which is in an oppositely directed vertical configuration. Circular motions perturb the unstable convective layer. The magnetic field is then dynamically forced into the convective downflow region, forming a current sheet in which the current density increases. As a result, magnetic reconnection is triggered. This process generates upward-propagating slow-mode MHD waves as well as Alfvén waves, provided a transverse magnetic field component is also present. These simulations showed that photospheric magnetic reconnection can produce solar spicules with sufficient energy at least through the lower chromosphere.

Two-dimensional simulations that include more of the relevant physics, including radiation and heat conduction, have been presented by Heggland *et al.* (2009). Waves propagating upward from the convection/photospheric zone were shown to induce reconnection events within a complex magnetic field topology. The simulations exhibit spicule-like jets with lengths and lifetimes that match observations, as well as reproducing some of their spectral signatures. Overall, spicule simulations demonstrate that there is a wide choice in the properties of the field shear and the field configuration. Together, they determine the nature of the jet flow, which is not surprising.

9.2.2 Coronal jets

All types of solar jet may share the same basic magnetic field topology that results in a collimated flow being driven along a large-scale unipolar guide field that reaches high into the corona (Moore *et al.*, 2010). The energy source of each type is rooted in an inclusion of opposite-polarity magnetic flux at the base of the jet. The opposite-polarity flux is associated with one foot of a small magnetic arch that has emerged or is still emerging from below the photosphere. The interface between the unipolar ambient field and the opposite-polarity leg of the arch leads to an inverted-Y or Eiffel-tower shape of many jets where abrupt reconnection occurs. As a new arch (or bipole) emerges within a locally unipolar magnetic field region, a neutral point appears immediately above it and magnetic reconnection occurs. The release of magnetic energy rapidly heats the plasma, which is then driven along both open fields and closed reconnected fields (Yokoyama & Shibata, 1995). The open field may only be locally open but can be just a small part near the footpoint of a much larger-scale loop.

Alternatively, an X-ray jet can be the 'evaporation flow' produced by a flare near the footpoint of a large-scale loop which follows a reconnection event (Shimojo & Shibata, 2000; Shimojo *et al.*, 2001). The model scenario begins with an abrupt deposition of thermal energy in the corona which is rapidly transported to the chromosphere by conduction. The dense plasma in the chromosphere is heated, the gas pressure rises, and a strong upflow of dense, hot plasma is ejected along the magnetic loop. The model, shown schematically in Fig. 9.6, is motivated by the close association of X-ray jets with flares at their footpoints. In addition, the magnetic reconnection model is also supported by the multipolar distribution of the magnetic field usually found, which is consistent with the magnetic reconnection model.

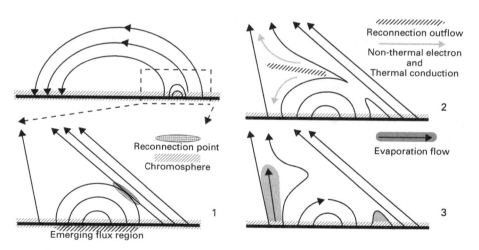

Fig. 9.6 Schematic diagram of the 'evaporation flow' model for X-ray jets as presented by Shimojo *et al.* (2001). The top-left sketch displays the entire region while the lower-left sketch zooms in on the emerging flux region. The creation of hot plasma is indicated at the top right. This plasma heats the denser layer near the footpoint via conduction (lower right), generating the jet. Reproduced by permission of the AAS.

Many of the active regions in coronal holes exhibit a structure that resembles a sea anemone (Asai *et al.*, 2008), characterised by radially aligned coronal loops that connect the opposite magnetic polarity of the magnetic field in the surrounding region with the unipolar field. In reconnection simulations, the classical anemone topology includes a null-point which is a strongly preferred site for reconnection and, hence, for jets to occur. It is clearly important to consider a three-dimensional magnetic null-point topology, since the reconnecting process is much more complicated than can be represented by the two-dimensional case. In this situation, the null lies between open and closed magnetic field lines. Reconnection slings cool plasma up along the open magnetic field and heats some plasma via shocks and resistive dissipation to extreme ultraviolet and X-ray temperatures.

There may be just two distinct types of observed jet flow, as discussed in Section 8.4. The non-standard jets are interpreted by Moore *et al.* (2010) as counterparts of erupting-loop macrospicules in which the jet-base magnetic arch undergoes a miniature version of the blowout eruptions that produce major coronal mass ejections. In this picture, the dichotomy depends on whether or not the base arches undergo sufficient shear and twist to cause them to erupt open.

9.3   Magnetic field methods

9.3.1   Hydromagnetic driving from rotating discs

The most advanced scenarios for launching jets from most astrophysical objects including quasars and young stars involve the rotation of large-scale magnetic fields. Not only the launch but the action of magnetic fields are attributed to all three key aspects of jet formation: ejection, acceleration and collimation. The specific models and mechanisms have been subject to much speculation. In general support, the observations inform us that magnetic fields are indeed present. Fields are detected indirectly and their strengths are constrained through synchrotron radiation (such as the radio emission of extragalactic jets) and in some cases measured through the Zeeman effect in spectral lines (OH or H_2O masers) in protoplanetary nebulae and young stellar objects. However, to establish the magnetic model will require more direct evidence. An unequivocal observational test of the correct magnetic field strength and configuration in the inner region where the flow is generated has proved to be elusive.

The magnetic field is taken to be threaded into a rapidly rotating object. The object could be a magnetically active star, a young pulsar or an accretion disc associated with a young stellar object, X-ray binary or active galactic nucleus. We first consider configurations in which the field is rooted in an accretion disc.

A general scenario has emerged, forged on many analytical and numerical studies. On the analytical front, disc winds have been extensively studied by taking various approximations. These include, amongst many others, assumptions of self-similarity in two-dimensional axisymmetric, time-independent flows (Blandford & Payne, 1982; Contopoulos, 1994), consideration of the asymptotic properties of the collimation (Heyvaerts & Norman, 1989) and permitting arbitrary variations of the gas pressure (Tsinganos & Trussoni, 1991; Trussoni *et al.*, 1997).

Numerically, the first revealing time-dependent ideal magnetohydrodynamic simulations were undertaken by Uchida and Shibata (1984) and then applied to accretion discs (Uchida & Shibata, 1985). The problem set-up in this case showed that the driving mechanism is analogous to an uncoiling spring as magnetic pressure from a toroidal field expels the gas. To generate the high magnetic twist, the disc is assumed to undergo periodic rapid radial collapse through instability or sub-Keplerian rotation rates. The accumulated magnetic twist in the disc eventually relaxes through the release of torsional Alfvén waves. The waves take with them the matter from the surface layer of the disc and eject it episodically. On the other hand, simulations of the centrifugally driven wind model take the disc to be in Keplerian equilibrium, imposed as a boundary condition at the base of the wind. The flow here is analogous to beads being flung out on a fast-rotating stiff bent wire. The gas is centrifugally accelerated through the Alfvén surface and the fast magnetosonic surface before being collimated into cylinders parallel to the disc's axis. The collimation is due to the pinch force exerted by the dominant toroidal magnetic field generated by the outflow itself (Ouyed & Pudritz, 1997).

The concepts of the standard magnetocentrifugal acceleration model (Blandford & Payne, 1982) can be understood by splitting the engine into three distinct physical regions separated by two transitional flow regions. The accretion disc supplies the mass and generates the forces on the magnetic field. In the disc, the rotational kinetic energy of the dense gas dominates over magnetic energy. Therefore, the lines of magnetic field, assumed frozen-in, are anchored onto and are forced to co-rotate with the disc.

The disc atmosphere is the first transition region in which a fraction of the disc material is transported. The rate at which material is lifted from the disc is very uncertain and is often entered as a free parameter, although it is suspected that 10% of the accreting material is involved (Pelletier & Pudritz, 1992). In the simplest form, the material would be raised by thermal pressure gradients supported by disc and external radiative heating to replace that lost through the magnetic torques. This needs to be only a small fraction of the accreting material. The flow is accelerated through the sonic point, corresponding to the slow magnetosonic speed, while still close to the disc surface. Given the density ρ at this point, the mass flux can be estimated as ρc_s, given the high value of v_a/c_s in a magnetically dominated atmosphere.

This flow enters the second region, the relatively low-density extended atmosphere on either side of the disc. Provided it is not too hot, the thermal pressure is low and the dynamics is controlled by the magnetic pressure. As in the solar atmosphere, the magnetic field is force-free. Any gas is thus forced to co-rotate with the underlying disc. Along the field, however, there is no restriction but the gas is subject to a strong centrifugal force. Hence, due to the frozen-in property of ideal MHD flows, in steady state, the flow will be along the field lines. As described above, we can picture one element of gas flowing along a field line as a 'bead' sliding on a rigid 'wire', or as being flung out within bent tubes of magnetic flux. The speed can be shown to increase almost linearly with distance from the rotation axis. The poloidal field in this region will be bent towards the rotation axis with increasing distance as the magnetic tension balances the magnetic pressure gradient.

A bead of gas may in fact be trapped or expelled, depending on the field geometry. If the rigid field lines are far from the vertical, the centrifugal force exerted on a bead of gas can compete against the gravitational force pushing the bead inward. Therefore, the acceleration depends on the inclination of the field lines, which must exceed some critical angle relative to the spin axis of the disc in order to be propelled outwards by the centrifugal force. This leads to an issue concerning collimation: to extract a high-speed flow requires a low collimation. Therefore, a subsequent distinct collimation region must be invoked.

A transition to a third region takes place as the flux tubes rapidly expand and the field strength decreases significantly with distance. Beyond some distance termed the Alfvén radius, the flow speed exceeds the Alfvén speed. The gas dynamic pressure then takes over from the magnetic pressure. The passive rigid rotation is replaced by inertial effects and the magnetic field begins to lag behind the disc rotation and so winds up into a toroidal configuration. This gives rise to magnetic hoop stresses which enhance the flow collimation. At this stage, the flow has already attained a significant fraction of the terminal jet speed.

In the third region, the field is essentially azimuthal as it continues to wind up. Here, other dynamical effects become important which may drive the outflow more effectively: a vertical magnetic pressure and reconnection.

In summary, the actual launch region can be considered as the transition region between regions 1 and 2. Mass from the high-β disc must be continuously loaded into the low-β atmosphere (where the plasma β is the ratio of thermal to magnetic pressure). The acceleration is largely achieved in region 2 and the second transition region, while the collimation takes place in region 3.

For acceleration of cold gas to occur, the magnetic field must be at an inclination of under 60° to the disc surface after the initial sonic transition has been made. To show this, we consider a cylindrical coordinate system with the rotating disc occupying the $z = 0$ plane, supported by the gravity from a central mass, M. We consider a parcel of gas at (r, z) which originates from $(r_o, 0)$ corresponding to a disc location at the radius r_o. It is constrained by the strong field to rotate with a constant angular frequency Ω which is equal to the Keplerian angular velocity at r_o: $\Omega^2 = GM / r_o^3$. Hence, the rotation speed is $r\Omega$ and the gas parcel possesses a specific kinetic energy $r^2 \Omega^2 / 2$. The gravitational potential energy is $GM / (r^2 + z^2)^{1/2}$. Therefore, the distribution of total potential energy is

$$\phi(r, z) = -\frac{GM}{r_o} \left[\frac{1}{2} \left(\frac{r}{r_o} \right)^2 + \frac{r_o}{(r^2 + z^2)^{1/2}} \right]. \tag{9.8}$$

The surfaces of constant ϕ can be drawn and, by (twice) differentiation, it is straightforward to show that the potential energy will fall as the gas leaves the location (r_o, z) along any direction with $(dz/dr)^2 < 3$. Hence, the equilibrium is unstable if the projection of the holding wire on the meridional plane makes an angle of less than 60° with the equatorial plane: for poloidal fields more vertical than this the balance is stable. Including the gas pressure will increase the angle (Pelletier & Pudritz, 1992). Consequently, with an oblique field, gas can easily be thrown out with only a small effort (due to the fact that the accretion disc material is in a state close to that of free fall, needing only a small push to escape).

It may still be possible to drive jets for near-vertical fields with inclinations of less than 30° to the axis. In this case, disc coronal plasma cannot be expelled by centrifugal forces alone but would require the assistance of strong thermal or radiative pressure. Indeed, radiation can be particularly efficient in powerful quasars that radiate at a significant fraction of the Eddington limit.

9.3.2   *Magnetocentrifugal quantities*

The purpose here is to determine fundamental properties of the magnetocentrifugal slingshot mechanism. We first demonstrate that a large fraction of angular momentum can be extracted from an accretion disc by ejecting a small fraction of the mass. We assume a stationary axisymmetric flow in a cylindrical coordinate system (r, ϕ, z). Due to the axisymmetry, the magnetic field can be decomposed into poloidal, $\mathbf{B}_p = \mathbf{B}_r + \mathbf{B}_z$ and azimuthal, $\mathbf{B}_\phi$, components. Also the velocity has a poloidal ($\mathbf{v}_p$) and a toroidal ($\mathbf{v}_\phi = \Omega \times \mathbf{r}$) component.

Mass and magnetic flux along field lines are both conserved quantities in this time-independent set-up. As we will see below, the degree of mass loading from the disc into the wind is crucial. We take a parameter k to be the ratio of outflowing mass flux, $d\dot{M}_{out}$, to threading magnetic flux, $d\Phi$, through an annulus $2\pi r_o dr_o$ of the disc of radial width dr_o. This mass load parameter, k, is thus only a function of the radial footprint distance in the accretion disc, r_o, from which the field line originates. The MHD equations can then be shown to yield

$$\mathbf{v}(r, z) = \frac{k\mathbf{B}(r, z)}{\rho} + (\Omega_\mathbf{o} \times \mathbf{r}), \tag{9.9}$$

where $\Omega_\mathbf{o}$ is the angular velocity of the disc at the disc mid-plane. That is,

$$\rho\mathbf{v}_p = k\mathbf{B}_p, \tag{9.10}$$

where k can be written as

$$k(r_o) = \frac{\rho v_p}{B_p} = \frac{d\dot{M}_{out}}{d\Phi}. \tag{9.11}$$

The toroidal field is generated within the wind via the rotation. The induction equation yields

$$B_\phi = \frac{\rho}{k}\left(v_\phi - \Omega_o r\right). \tag{9.12}$$

This demonstrates how the toroidal field forms by the winding up of the initial polar field from the disc surface.

The total angular momentum per unit mass is also a conserved quantity. It includes the angular momentum associated with the coiled-up field lines, which, if released, would spin up the rotating flow:

$$l(r_o) = rv_\phi - \frac{rB_\phi}{4\pi k} \tag{9.13}$$

(Pelletier & Pudritz, 1992). Note that a higher mass load reduces the relative angular momentum of the magnetic field. This equation is instructive by showing how the toroidal magnetic field increases as the angular momentum of the gas increases, so keeping the total angular momentum constant.

As noted above, the Alfvén point signifies the scale of the flow out to which the magnetic field lines act as rigid wires. We now define the Alfvén point and (cylindrical) radius, r_A, as where the poloidal speed reaches the speed of poloidal magnetic waves, i.e. the poloidal Alfvén speed. At this location, the Alfvén Mach number,

$$M_{A,p}^2 \equiv \frac{4\pi\rho v_p^2}{B_p^2}, \tag{9.14}$$

reaches unity. Also at this point, the specific angular momentum takes the value

$$l = \Omega_o r_A^2 = (r_A/r_o)^2 \Omega_o r_o^2, \tag{9.15}$$

a factor $(r_A/r_o)^2$ larger than when the material left the disc (Pelletier & Pudritz, 1992). Therefore, the magnetic field acts as a lever arm to extract a much larger portion of the disc's angular momentum than contained in the fraction of mass ejected. Moreover, it is feasible to consider that the wind carries away all the disc angular momentum in order for accretion to take place. This obviates the need for angular momentum transport by viscous torques in the disc. Furthermore, such a disc would be silent since it would also extract the entire accretion energy.

Other significant flow quantities have been estimated by Pelletier & Pudritz (1992) for the case of cold flows, as follows. The terminal speed is

$$v_f \sim 2^{1/2}\Omega_o r_A, \tag{9.16}$$

which depends strongly on the footpoint radius from which the flow begins. This shows that the ensuing jet speed on large scales will be strongly sheared with high axial speeds. However, speeds are well in excess of the local escape speed due to the effect of the magnetic lever arm.

If the angular momentum from the disc is extracted exclusively through the outflow, then the mass outflow rate is estimated as

$$\dot{M}_{out} \sim (r_o/r_A)^2 \dot{M}_{accr}, \tag{9.17}$$

where $\dot{M}_{accr}$ is the accretion rate. Therefore, as an example, a lever arm ratio of 3 would imply that 11% of the accreting mass is ejected.

Most of the outflowing power is confined to an axial core flow. In this sense, the jet can be considered to be the dense core of a wind along the rotational axis. The total power can be estimated from Eq. (9.16) and (9.17). On integrating over the entire disc, one finds

$$P_{jet} = \frac{GM \dot{M}_{accr}}{2 r_{inner}}, \tag{9.18}$$

where r_{inner} is the inner disc radius. Remarkably, the mechanical power in the jets is half the accretion luminosity. Thus, the wind taps the release of gravitational energy which is required in accretion models.

Many variants of and extensions to the disc-launched magneto-centrifugal model can be found in the literature. Contopoulos and Lovelace (1994) introduced a relativistic version and discussed configurations where an electric current flows along the jet, causing the jet radius to oscillate with axial distance. The distribution of the magnetic field across the disc is also a vital factor (Anderson *et al.*, 2005; Fendt, 2006), as is the mass loading at the base of the wind (Ouyed & Pudritz, 1999).

9.3.3   X-winds

Numerous launch scenarios are based on the interaction of a central object with a viscous and imperfectly conducting accretion disc. The theory is based on either solving the equations of Section 3.5, especially the Grad–Shafranov equation, or solving numerically a set of MHD equations with resistivity.

The original extraordinary-wind model, as shown in Fig. 9.5, assumed that the accretion disc abutted the surface of a magnetised star (Shu *et al.*, 1988). The wind then arises from the equator as the star is spun up to break-up. Later, however, it was realised that classical

T Tauri stars are only rotating at one-tenth of the break-up speed. Therefore, a configuration in which the rotating magnetosphere of the star truncates the accretion disc, with a relatively low accretion rate, needs to be considered.

The resulting X-wind model embeds an alternative magnetocentrifugal mechanism for jets (Shu *et al.*, 1994) as part of a more general scheme for the evolution of young stars. In this picture, the stellar magnetosphere is represented by a globally closed dipolar field configuration. The stellar field lines come up against the disc but are stopped from squeezing it by shielding currents on the disc surface which yield a disc field. Therefore, the stellar field is only capable of penetrating the disc over a small range of radii about an inner radius r_x, where the Keplerian angular speed of rotation Ω_x equals the angular speed of the star Ω_*.

However, if most of the mass of the central star is still to be accumulated through the disc, the magnetic coupling between the star and the disc forces the star to co-rotate at the Keplerian frequency associated with the truncation radius. For low disc accretion rates and high magnetic field strengths, likely to be associated with classical T Tauri stars, R_x exceeds the radius of the star R_* by a factor of a few (Ostriker & Shu, 1995).

Most of the material is funnelled off from the accreting, electrically conducting disc onto the stellar surface. In detail, the inner disc is effectively truncated at a radius R_t somewhat smaller than R_x. This so-called X-region is small, with a fractional size thought to be given by the ratio of disc surface sound speed to the local Keplerian speed where the X-wind is launched (Shu *et al.*, 2007). The closed stellar field lines crossing the inner region between R_t and R_x bow inwards. The accreting gas attaches itself to the stellar field and is funnelled down the effective potential (gravitational plus centrifugal) onto the star. Thus, the inner edge of the disc is forced to co-rotate with the star. The star is not spun up at this stage because of magnetic torques associated with the accreting gas which offload angular momentum to the associated wind rather than to the star.

Some disc material will be ejected in a wind. The magnetic field lines threading the disc exterior to R_x will bow outward as shown in Fig. 9.7. The disc gas off the mid-plane then rotates at super-Keplerian velocities. This drives a magnetocentrifugal wind with a mass-loss rate equal to a definite fraction of the disc accretion rate, which is typically between 1/4 and 1/3, the rest being funnelled in to the star.

An alternative model to the X-wind may be realised when the star is a rapid rotator so that the magnetospheric boundary rotates at super-Keplerian velocities (Romanova *et al.*, 2009) or where the inner radius of the disc exceeds the co-rotation radius (Ustyugova *et al.*, 2006). The interaction of a large amount of accreting material with this boundary is analogous to feeding a fan – a propeller effect – with the ejection of almost all the disc mass. This mechanism may be appropriate for jets driven by white dwarfs and neutron stars.

The X-wind is generated by the magnetocentrifugal slingshot mechanism. At the truncation radius, the equatorial inward drift within the accretion disc results in an oblique angle between the stellar magnetic field lines and the disc normal. If approximate field freezing holds as the accretion proceeds, some of these field lines will be bent through an angle exceeding 30°. Hence, matter frozen to these flux tubes becomes unstable (Blandford & Payne, 1982). These field lines are thus responsible for driving an MHD wind from the disc. Since the wind removes angular momentum, the footpoints of those field lines in the disc will try to migrate inward. The radially inward press of these footpoints and the

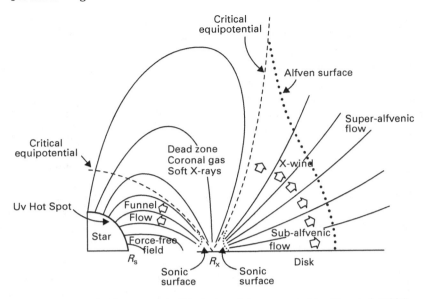

Fig. 9.7 Schematic diagram of the X-wind model as presented by Shu *et al.* (1994). Reproduced by permission of the AAS.

radially outward press of the footpoints of the funnel-flow field lines together create a magnetic X-configuration that distinguishes the model from similar variants in the literature. A quasi-steady state in envisaged in which the radial advection into the X-point is balanced by the resistive diffusion of field lines out of the X zone. However, the interface between open field lines loaded with outflowing matter from the disc and those from the star forms a configuration called a 'helmet streamer' (Ostriker & Shu, 1995). This interface will be the origin of major mass-ejection and reconnection events possibly triggered by the amplified activity of pre-main-sequence stars.

On the larger scale, the outflow collimates into jets along the rotation axis. Shu *et al.* (1995) demonstrated that the wind density asymptotes to cylindrical contours while the streamlines become radial. This has led to some comparison to the observations including predictions for the thermal and ionisation properties (Shang *et al.*, 2002), leading to a calculation of the free-free radio emission (Shang *et al.*, 2004). For the HH 212 system, Lee *et al.* (2008) estimated the wind launching radius to be 0.05–0.30 AU, consistent with the X-wind or disc-wind models. Other simulations of optical forbidden line emission from jet shock waves lead to some extension in disc radii being favoured (Staff *et al.*, 2010). However, in the 'magnetized accretion ejection structures' model of Ferreira and Pelletier (1995) it has been cautioned that one has to consider the interdependence of accretion and ejection physics to obtain a realistic solution. Crucially, the collimation properties depend on the radial distribution of the mass loading into the wind (Ferreira, 1997).

Jet flows may be inherently unsteady and unstable. A time-dependent jet launching and collimating mechanism was discovered in MHD simulations of the interaction between an aligned dipole rotator and a conducting circumstellar accretion disc by Goodson *et al.* (1997). The mechanism depends on the differential rotation between the star and the region of the disc to which it is magnetically connected. The outflow consists of two components with a hot, well-collimated axial jet separated from a cool, slow disc wind by rapidly expanding magnetic loops. The inflating loops are maintained through repeated magnetic reconnection events.

The operation of the X-wind does not require a dipolar geometry for the unperturbed stellar magnetosphere. In fact, accreting T Tauri stars have long been known to possess surface hot spots, which suggests a complex field structure. The superposition of a strong quadrupolar magnetic field may generate one-sided flip-flopping jets (Lovelace *et al.*, 2010). The generalized X-wind picture developed to account for stellar activity shows that the outflow from a small annulus near the inner disc edge is hardly influenced by a modified geometry (Mohanty & Shu, 2008). However, the funnel flow is radically altered by the interaction of the emergent stellar flux with the inner disc region.

Relativistic MHD winds driven by rapid rotators were treated by Camenzind (1987). In this scenario, accreting matter is injected into the magnetosphere of a rapidly rotating compact object. This requires the inner edge of the accretion disc to be located between the co-rotation radius and the light-cylinder radius. Results were obtained by solving the relativistic Grad–Schlueter–Shafranov equation with a finite-element method. The resulting poloidal wind speed can be a significant fraction of the speed of light if the asymptotic Poynting flux approximates to the total energy flux. The Lorentz factor reached is in the range of 1 to 5, which suggests a potential interpretation of quasar jets.

9.3.4 Spinning black holes

The properties of a black hole are entirely described by just three parameters: mass, charge and angular momentum. Some of the mass is actually reducible – that is, the equivalent energy can be extracted through classical processes. This is impossible for a static (Schwarzschild metric) black hole with no rotation or electrical charge, which has a 100% irreducible mass (energy can then only be extracted through the Hawking evaporation process of quantum physics). However, a rotating (Kerr metric) black hole will possess reducible mass (negative energy can be accreted when the black hole rotates). The rotational energy as well as the Coulomb energy are indeed extractable by physical means. In the present context, the Blandford–Znajek mechanism provides the framework for the extraction in which the energy may be converted into jets (Blandford & Znajek, 1977).

The mechanism supposes that a rotating black hole accretes material from a magnetised accretion disc. The frozen-in disc field increases in strength as it approaches the event horizon of the black hole. The field is supported by strong electric currents flowing in the disc, while the disc can be Keplerian (maintained by centrifugal force) or magnetically supported. In this model, the disc serves only to carry in and maintain the magnetic field in the black hole ergosphere.

An electric field is induced in the vicinity of the horizon which accelerates any stray particles along field lines. If the field and angular momentum are sufficiently high, the acceleration generates radiation which in turn generates further electron–positron pairs, leading to a cascade. The charges ensure that the electromagnetic field becomes approximately force-free. The charged plasma has negligible inertia and plays no other role in the scenario apart from supporting the currents associated with the magnetic field and cancelling the electric field in local fluid frames.

The electromagnetic field, however, will transport energy and angular momentum – an outward electromagnetic energy flux occurs from the event horizon. The extraction requires a stress through which the black hole does work on the field. For the flux to be outgoing, the field lines must be rotating slower than the black hole. Hence, a load is envisaged that inhibits the field lines from spinning up. The Poynting flux is then given in terms of the strength of the magnetic field and its rotation rate.

How much of and how fast will the energy be extracted from the black hole? There is no definitive answer, it being dependent on several factors. The energy stored may be extremely large. To show this, we take a black hole of mass M to possess angular momentum $J = aMc$, where a is a length called the spin parameter which is restricted to the range $0 < a < GM/c^2$. The reducible spin energy is $E_{spin} = Mc^2 - M_{irr}c^2$ where M_{irr} is the irreducible portion. For simplicity, we transform into units such that $G = c = 1$. We then obtain that $0 < a < M$ and

$$M_{irr}^2 = \frac{1}{2}Mr_+,$$ (9.19)

where the outer event horizon is given by

$$r_+ = M\left[1 + \left(1 - (a/M)^2\right)^{1/2}\right]$$ (9.20)

(e.g. McKinney & Gammie, 2004). For a maximally rotating hole this yields $E_{spin} = 5.3 \times 10^{61}(M/10^8 M_\odot)$ erg which is just under 30% of the gravitational mass. Even if this energy is extracted over an entire Hubble time, the luminosity would be $4 \times 10^{10}(M/10^8 M_\odot) L_\odot$. Nevertheless, it has been found that the jet luminosity will still be limited to the accretion power associated with the rest-mass and internal energy of the accreting plasma (Ghosh & Abramowicz, 1997). A portion of the mass-energy is radiated away during the accretion and the rest is incorporated into the black hole. This energy then gets a second chance to escape through electromagnetic spin-down.

A further issue is how fast a black hole is spun up in practice, which depends strongly on how the accretion connects to the black hole (De Villiers *et al.*, 2003). Models based on simulations have identified four regions: a force-free funnel region, a corona, an equatorial disc, and a plunging region between the last stable orbit of the disc and the event horizon. Thus, due to the nature of the latter, the black hole spin can influence the accretion rate.

9.3.5   *Poynting jets*

The average particle energy far exceeds the rest mass energy in many observed jets, especially those associated with blazars and gamma-ray bursts. Therefore, a large fraction of the available energy must be transferred into a small amount of matter. This can be achieved if the energy is initially electromagnetic and transported in the form of a Poynting flux in which the magnetic energy density far exceeds the plasma energy density. However, if a jet is indeed accelerated by magnetic fields in a Poynting-dominated regime but terminates as a matter-dominated flow, a transition zone and process must be identified.

The generation of rotating, twisted magnetic fields is indeed the basis for the recent launching models of powerful jets from black holes, neutron stars and accretion discs. The gas flow patterns have been described in terms of ideal magnetohydrodynamics, including the Lorentz force even though force-free solutions may be appropriate throughout the launch zone in the relativistic case. Nevertheless, the force-free flow can usually be considered as the limiting case in which the particle mass and pressure are negligible. This is the basis of Poynting flux-dominated jets which can exist in both a non-relativistic (Kudoh & Shibata, 1997) and a relativistic (Vlahakis & Königl, 2001) form.

It should be noted that the creation of a jet with a highly relativistic speed can, in principle, be achieved hydrodynamically. If the plasma has an initial high pressure, a relativistic form of the de Laval nozzle will occur. Along the flow, thermal energy is converted into the bulk kinetic energy according to the Bernoulli equation. Assuming a relativistically hot fluid with the adiabatic index of 4/3, the flow Lorentz factor grows in proportion to the jet radius (Lyubarsky, 2011). The principle, however, requires conditions that are not found: a highly relativistic fluid source that behaves adiabatically within a high-pressure confining medium. Detailed calculations place strong doubts on hydrodynamic driving of gamma-ray bursts (Daigne & Mochkovitch, 2002) and active galactic nuclei (Vlahakis & Königl, 2004).

The pressing issue of much research is how the electromagnetic energy can eventually be transferred quite efficiently into the plasma. The field may be initially poloidal (corresponding to trans-Alfvénic solutions), but a toroidal component becomes dominant on larger scales and the plasma is ejected by the resulting magnetic tension. In the magneto-centrifugally driven model, the gas is already strongly accelerated at the base of the flow along the rotating poloidal field lines. The toroidal field is generated only when the gas inertia becomes comparable to the magnetic stresses. The inertia causes the field lines to bend backwards. This implies that a good fraction of the Poynting flux has already been converted into the kinetic energy of the flow before the toroidal field has become comparable to the poloidal one.

Poynting jets or *Poynting flux-dominated* jets contrast with this. The poloidal magnetic force is not balanced by inertia but by the electric force. Hence these are termed *current-carrying jets*. The toroidal field becomes comparable with both the poloidal field and the electric field at the light cylinder, the surface on which the co-rotational speed is equal to the speed of light. Interior to this surface, the magnetosphere co-rotates with the central object and the fluid slides along the rotating field lines. On reaching the surface, the fluid kinetic energy is still very small (Lyubarsky, 2009). Beyond the light cylinder, however, the poloidal magnetic field decreases relatively fast, $1/R_{jet}^2$, with cylindrical radius R_{jet}, according to the conservation of magnetic flux. The azimuthal field B_ϕ and the electric field then conspire to reach a close force balance, both decreasing as $1/R$. Hence, there is only a small residual force on the fluid (Begelman & Li, 1994). In fact, the variation of the quantity $B_\phi R_{jet}$ is crucial to the rate at which Poynting flux is converted into kinetic energy. The result is that the fluid acceleration is extended well beyond the light cylinder within a Poynting-dominated funnel jet. The jet acceleration may thus span a very large range of scales (Lyubarsky, 2009) with final jet Lorentz factors exceeding 100.

The conditions for collimation and acceleration of relativistic magnetohydrodynamic jets and the efficiency of conversion of the Poynting flux are now being thoroughly studied analytically and numerically by many authors (e.g. Vlahakis & Königl, 2003). The mechanism is distinct from that of centrifugal acceleration, involving acceleration by the pressure gradient of the azimuthal magnetic field. By extending the radially self-similar MHD flows of Blandford and Payne (1982) to the relativistic regime, Li *et al.* (1992) showed that the magnetic pressure gradient is effective beyond the classical fast-magnetosonic point. The slight divergence of flux tubes then leads to effective fluid acceleration via a magnetic nozzle effect. The acceleration continues over an extended region up to a modified fast-magnetosonic surface. Jets dominated by Poynting flux may well be stable to current-driven z-pinch instabilities (see Section 10.5). However, these models are still under development, and it is not clear if stable, long-term collimated outflows over long distances can be reproduced.

9.4   Alternative models and mechanisms

9.4.1   *MHD simulations*

Building on earlier ideal magnetohydrodynamic simulations of Uchida and Shibata (1984), a reconnection model was advanced by Hirose *et al.* (1997) in which both the disc and star anchor magnetic fields, as shown in Fig. 9.8. The disc field, carried inwards in the accretion process, is assumed to be in the same direction as that of the star. This results in a magnetically neutral ring in the equatorial plane near the truncation radius. It was shown that reconnection then leads to the channelling of over 90% of the material into the magnetosphere while 10% escapes along open magnetic field lines, accelerated by the Lorentz force. Further acceleration to escape speeds was hypothesised to take place through the magneto-centrifugal mechanism, but not actually simulated.

9.4.2   *Magnetic towers and funnels*

A magnetic tower has a strong, tightly wound helical magnetic field as the defining feature. The tower principle is based on the theorem of Lynden-Bell and Boily (1994) which states that the force-free magnetic configurations expands along a cone of semi-angle 60° before reaching a flashpoint where a toroidal field is released vertically. A confining pressure which decreases rapidly with height may help accelerate and collimate the flow, in addition

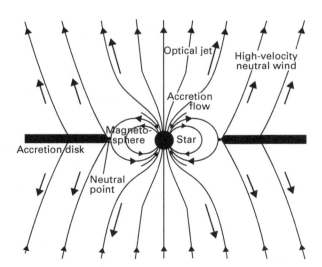

Fig. 9.8 Schematic diagram of the dual magnetised disc and stellar magnetosphere model as presented by Hirose *et al.* (1997).

to unwinding the field (Lynden-Bell, 2006). The configuration could be formed out of a large rotating torus which is threaded by a poloidal magnetic field. As torus material approaches the immediate environment of the central black hole, a magnetic jet emerges with a two-component structure. A strong axial poloidal field is surrounded by a toroidal field (Kato *et al.*, 2004). In the funnel model, the toroidal magnetic field as well as gas pressure together provide the collimating mechanism (Fukue *et al.*, 1991).

In another version, a tightly wound central helix propagates inside the tower jet, and a loosely wound helix returns at the outer edge of the magnetic tower jet (Nakamura *et al.*, 2006) A large forward current flows parallel to the jet axis and is responsible for generating the strong inner helical field. The return current is conically shaped and flows outside. This is part of a global picture in which a current-carrying jet possesses a closed current system along the lines introduced by Benford (1978) and applied to extragalactic radio sources. This closed current system consists of a pair of current circuits, each containing forward and return electric current paths.

9.4.3 *ADAF jets: ion-supported flows*

A cold disc which is optically thick but geometrically thin provides the landscape and launch material for jets associated with X-ray binaries observed in the high/soft state and luminous quasars with optical/ultraviolet spectral bumps. Black-body emission from the disc can account for the emission properties. Contrastingly, a hot, optically thin, geometrically thick accretion flow model has been developed to interpret low-luminosity active nuclei and the low/hard state of X-ray binaries (Ichimaru, 1977). Such flows are radiatively inefficient or advection-dominated accretion flows: RIAFs and ADAFs . For galactic nuclei, changing the physical state of the accretion flow may provide the switch from the radio-loud to the radio-quiet sources, while for X-ray binaries, the change in the nature of the accretion would arise at about 1% of the Eddington luminosity.

The ADAF took the form of ion-supported tori in the model of Rees *et al.* (1982) for the origin of radio jets. The ion torus acts to anchor the magnetic field which extracts the rotational energy from the black hole in the form of jets. The power created by accreting matter can either be radiated away, advected into the black hole, or taken away by jets. In the ADAF state, the accretion power is reduced as a result of reduced radiative efficiency and the jet power dominates below some critical accretion rate. These strong-jet low-powered sources were identified as FR I galaxies and low/hard-state X-ray binaries by Wu and Cao (2008).

9.4.4 *The disc–jet connection*

An alternative model is based on a switch between two states, as proposed by Livio *et al.* (2003). Here, relativistic jets are directly powered by accretion in one state. In the other state, the accretion energy is dissipated locally within the disc and associated corona, producing the observed high disc luminosity. The switch itself is provided by the generation of a global, poloidal magnetic field. This model was inspired by the observations of GRS 1915+105 (Section 7.5.3) for which the low/hard states are quite prolonged, implying that the disc cannot rapidly rebuild (Belloni *et al.*, 1997). In response, Livio *et al.* (2003) proposed that the inner accretion disc may actually still be present in the low state and processing material at essentially the same rate as in the high state. However, the liberated energy in the accretion is efficiently converted into magnetic energy and disposed of through magnetically-dominated jets. As the accretion state changes, the jets become visible through pulsations in velocity associated with the generation of a poloidal magnetic field. The accretion energy goes directly from velocity shear into toroidal and poloidal field loops, amplified with the aid of the magneto-rotational instability (Balbus & Hawley, 1991). An inverse cascade of random small-scale field loops generates a poloidal field along which material can be accelerated.

9.4.5   *Radiation-driven jets*

Radiation pressure is generally not sufficient to drive powerful jets (Phinney, 1982). In principle, radiation pressure on spectral lines and dust may drive a relativistic wind. The mass loss is supplied by the accretion disc and the radiative thrust from the central source. If the driven gas is optically thick, however, the maximum speed achievable is the relativistic sound speed, $c\sqrt{3}$.

Once a flow has reached a mildly relativistic speed, radiation becomes an ineffective acceleration mechanism. In fact, isotropic radiation acts as a source of drag with deceleration proportional to Γ^2 in the flow as a consequence of Compton drag. Except for the radiation which is directed very close to parallel to a blob, the photons will actually gain momentum due to the interaction (Phinney, 1982). Hence, a blob will slow down – radiation is an effective deceleration mechanism. In fact, Compton drag imposes severe constraints on any model of jet acceleration (Fukue, 2000) and some derived Lorentz factors of blazar jets exceed the predicted upper bounds of Compton-drag theories.

9.5   Summary

The most applicable jet paradigms invoke magnetic driving from either magnetised accretion discs or accreting, rotating black holes. The models involve the production of strong toroidal fields that form, accelerate and collimate the jets.

In light of the seminal work of Blandford and Payne (1982), it has been acknowledged that jets could extract a significant fraction of the underlying disc angular momentum and accretion power. Jets then take on a causal role by sustaining the accretion. Rather than being the smoke rising from the chimney, they help light the fire. Magnetic fields are critical to jet formation under the most extreme conditions. However, the existence or generation of a large-scale magnetic field in the inner regions of accretion flows is still an open and highly debated question.

For black holes, we have discussed most of the potential scenarios in detail. For normal stellar objects, the launching region must be quite close to the stellar surface in order to gather sufficient speed. The models can be classed by the location of the launching zone. The magnetic field lines are anchored at the stellar surface (e.g. Sauty *et al.*, 2002), at the interface between the stellar magnetosphere and the accretion disc (Shu *et al.*, 1994), or in the inner regions of the accretion disc (Pudritz *et al.*, 2006). Recent models attempt to accrue the advantageous aspects of at least two of the models (Matsakos *et al.*, 2009).

It remains difficult to relate the launch model to the observations even for cometary jets where we are closest to resolving the launch site. There are many assumptions and unknowns behind both the simulations and the analyses which allow for contradictory conclusions to be advanced. One means of progress is to consider the theory of jets as they propagate out to scales where they can be studied. For most classes of jets we are now able to penetrate down to within 100 times the size of the launch region. Therefore, we shall be able to confront theory with observation in the years ahead.

10

Jet propagation

The moment a jet has been launched and partly directed, its high thrust will vacate a channel. In an initial development stage, however, the impact region is close to the launch site. If the jet is maintained, the impact region recedes from the site using some of the jet momentum to drill the channel. A continuous supply of momentum will act to resist the ambient medium from re-entering the vacated space and so sustain a lengthening jet. Hence, the jet will become progressively longer provided the flow remains stable and fixed in direction. An early comprehensive review of the propagation physics for radio galaxies was provided by Begelman *et al.* (1984).

We reconsider here the conditions for a jet to remain supersonic, stable and well collimated out to large distances. We also consider the complex structures generated by dissipative and disruptive processes. A large part of our knowledge is derived from computer simulation, a method which has come a long way since the pioneering attempt by Rayburn (1977). The extensive literature necessitates a selective approach.

10.1 Components and structure

A basic description of a jet is provided by the opening angle. How the opening angle changes with distance from the source then describes the collimation properties. In theory, assuming a jet with a circular cross-section and sharp boundaries, the opening angle is well defined as $2 \times dr_s/dz$, where the jet boundary is given by by $r_s(z)$ in cylindrical coordinates. This may also be termed the local opening angle. However, quoted opening angles are sometimes simply the quantity $2 \times r_s(z)/z$.

Observationally, the apparent opening angle is a measure of the jet in projection onto the sky. This angle always exceeds the intrinsic value, owing to the foreshortening of the distance z due to the orientation. Taking the inclination angle, θ, between the jet axis and the line of sight, the measured jet length is $z_m = z \sin \theta$ and the apparent opening angle is $(2/\sin \theta) \times dr/dz$.

Observationally, however, we invariably have additional issues to contend with. Besides broadening due to the telescope beam and the sky seeing, the jet boundary depends on the radiation process and the strength of the radiation. The opening angle thus depends on the tracer being measured and the decreasing flux level. In practice, we measure the width of the jet as a function of distance, with the width defined as the full width at half maximum intensity or a similar quantity. To obtain the intrinsic width then requires a careful deconvolution to be applied to remove broadening due to the telescope, recording instrument and seeing.

Table 10.1. *The flow through jets: typical or conceivable Mach numbers, mass outflows and powers*

Jet type	Proper Mach number	Mass outflow rate rate ($M_\odot$/year)	Flow power (erg s^{-1})
Cometary	3	10^{-19}	10^{16}
Solar spicule	10		
Solar coronal	10		
Protostellar	100	10^{-5}	10^{35}
Herbig–Haro	30	10^{-7}	10^{33}
T Tauri microjet	30	10^{-8}	10^{32}
Planetary nebula	10	10^{-7}	10^{33}
Symbiotic star	60	10^{-6}	10^{36}
Supersoft source	100	10^{-7}	10^{36}
Cataclysmic var.	100	10^{-8}	10^{33}
Low-mass XRB	10	10^{-8}	10^{36}
High-mass XRB	10	10^{-6}	10^{40}
Microquasar	10	10^{-6}	10^{40}
Pulsar	10	10^{-11}	10^{31}
Gamma-ray burst	100	10^{1}	10^{50}
Blazar/quasar	10	10^{0}	10^{44}
Radio galaxy FRI	3	10^{0}	10^{40}
Radio galaxy FRII	10	10^{0}	10^{44}

In the classical jet theory, the jet is a well-defined single component. It is a supersonic flow presumed to be in contact with and pressure-confined by a stationary thermal medium. In modern theory, however, a collimated outflow has several components which can be identified in a global jet flow. The actual jet remains a specific term for the launched high-speed collimated flow, which must be distinguished from the sheath, shear layer, spine and surrounding cocoon.

The cocoon was uncovered in global simulations which illustrated that the jet vacates a cavity as it ploughs through the ambient medium (Fig. 10.1). The cavity is filled with the overspill from the jet resulting from the geometry of the impact zone where the jet material is shocked (Scheuer, 1974). The shocked gas is deflected laterally and accelerated back towards the source, thus forming an extended hot cocoon. The cocoon is particularly wide when the jet is of high Mach number and very light. If the ambient density, ρ_a, far exceeds the jet density, ρ_j, the shocked jet material is raised to a high pressure and must expand laterally. Hence, the jet actually interacts directly with a cocoon of hot ejected gas rather than a quiescent ambient medium. Furthermore, the cocoon may maintain an elevated pressure around the jet, especially in the vicinity of the impact zone, simply due to the comparatively shallow pressure gradient across the cocoon.

The impact zone itself is associated with hot spots in extragalactic sources (Section 4.2.2) and bow shocks or working surfaces (Section 6.3) in outflows from young stellar objects.

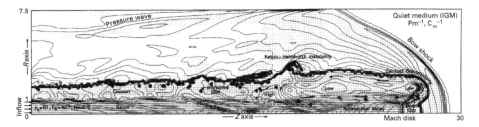

Fig. 10.1 A two-dimensional (cylindrical) simulation of a light supersonic jet ($\rho_j/\rho_a = 0.1$, $M = 6$). Adiabatic hydrodynamics is assumed. Shown here are density contours. Extracted from Norman *et al.* (1982).

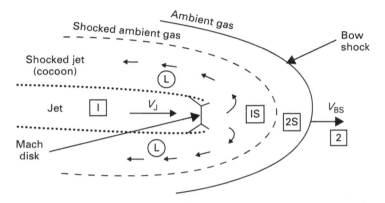

Fig. 10.2 A schematic picture of a jet-driven bow shock as drawn by Hartigan (1989). Reproduced by permission of the AAS.

The zone contains two major shock waves, a reverse shock in the jet and a forward shock into the ambient medium, as sketched in Fig. 10.2. The two shocked media are then separated by an interface which may be represented, in the two extremes, by a contact discontinuity or by a strongly unstable shear layer. In the hydrodynamic non-relativistic case, the advance speed of the impact zone, V_i, is given by equating momentum fluxes entering the impact zone:

$$\left[\rho_j(v_j - V_i)^2 + p_j\right]A_j = \left[\rho_a V_i^2 + p_a\right]A_b,\tag{10.1}$$

where A_j and A_b are the (effective) cross-sectional areas of the jet and bow shock (e.g. de Gouveia dal Pino & Benz, 1993). In this manner, the advance speed is estimated by equating the thrust of the jet into the impact zone to the thrust injected into the swept-up ambient gas. In a one-dimensional approximation, the bulk momentum flux per unit area reaching the hot spot is just the ram pressure $\rho_j(v_j - V_i)^2$, while the ram pressure on the ambient gas is $\rho_a V_i^2$. Equating ram pressures yields

$$V_i = \frac{v_j}{1 + 1/\sqrt{\eta}},\tag{10.2}$$

where $\eta = \rho_j/\rho_a$ is the density ratio. This formula is based on several assumptions. These include that the shocks are strong and non-relativistic, that the ambient gas is swept up rather than deflected out of the path, that the entire impact zone remains thin, and that the specific heat ratios are the same for both media. In the relativistic case, the same formula holds with

$$\eta = \frac{\rho_j h_j \Gamma_j^2}{\rho_a h_a}, \tag{10.3}$$

where h is the specific enthalpy and Γ_j is the Lorentz factor for the jet (measured in the rest frame of the ambient medium) (Marti *et al.*, 1997). As noted by Marti *et al.* (1997), the inertial mass density has contributions from both the internal energy and the Lorentz factor in relativistic dynamics. Therefore, for the same parameters, relativistic jets tend to possess less prominent cocoons.

A light jet with $\eta \sim 0.1$ such as illustrated in Fig. 10.1 generates a cocoon which can be interpreted as the lobes of radio galaxies. However, the large cavities discussed recently (Section 4.2) can only be generated if the jet is considerably lighter: $\eta \sim 0.01$ or even lower (e.g. Krause, 2003).

While simulations of uniform jets within uniform media have revealed the basic physics, models of expanding very light jets propagating through non-uniform atmospheres should produce the global and evolutionary characteristics. Three stages were thus identified in the work of Reynolds *et al.* (2001). The jet begins by supplying an overpressured cocoon that drives a wide, strong shock into the intracluster medium as it expands supersonically. The cocoon pressure will decrease, and the shock strength with it. In this second regime, the ambient gas is mainly disturbed by the *sonic boom* of the jet impact. This stage corresponds to that identified by many Chandra observations of cool X-ray rims (Section 4.2.1). In the final stage, the cocoon collapses and the X-ray shell fades. Amongst numerous variations on this theme, a magnetically dominated version was presented by Xu *et al.* (2008) which demonstrated that the cavity can remain intact for periods exceeding $\sim 500\,\text{Myr}$.

A jet which is much denser than the ambient medium will drive through at a speed close to that of the jet speed, since $\eta \gg 1$. There is then only a small mass flux from the jet into the hot spot and the cocoon is narrow. This leads to so-called naked jet morphologies (Norman *et al.*, 1983) with the jet overspill forming a lobe adjacent to the hot spot. According to this theory, the shocks of the working surface of a dense jet are less dissipative than those of a light jet, since the jet motion is ballistic and the deceleration is minimal. To quantify, we note that the total energy dissipated per unit area is approximately

$$P = \frac{1}{2}\rho_j(U - v_j)^3 + \frac{1}{2}\rho_a U^3, \tag{10.4}$$

which can be rewritten (Rosen & Smith, 2004b) as

$$P = \frac{1}{(1 + \sqrt{\eta})^2} \times \frac{1}{2}\rho_j v_j^3. \tag{10.5}$$

This indicates that a steady dense jet cannot directly impart its energy into the ambient medium. Moreover, in the context of protostellar jets, collimated supersonic jets are only able to stir up the ambient medium in the immediate vicinity of the jet itself. As suspected, simulations have shown that supersonic turbulence decays rapidly and little energy can be transmitted spatially into the entire molecular cloud (Banerjee *et al.*, 2007).

The above formulae are based on one-dimensional ram pressure balance and define a benchmark for the propagation efficiency. These results have been somewhat modified

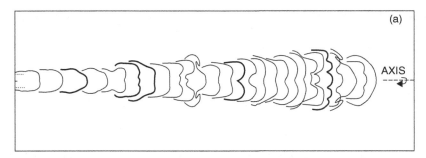

Fig. 10.3 The evolving interface structure from the $M = 6$ light jet simulation displayed in Fig. 10.1. The interface is displayed at equal time intervals. Taken from Smith *et al.* (1985).

through numerical simulations. It is found that the propagation efficiency depends not only on the geometric treatment but also on the physical treatment of the gas and magnetic field. The efficiency increases with jet Mach number for $M > 3$ (Norman *et al.*, 1983; Massaglia *et al.*, 1996).

The morphology of the hot spots produced in simulations of light adiabatic jets is constantly transforming Fig. 10.3, (Smith *et al.*, 1985; Saxton *et al.*, 2010). The time dependence is a consequence of the high pressure gradients set up by the toroidal vortex motion as jet material is shed from the jet and diverted into the backflowing cocoon. The vortex motion feeds back onto the jet gas approaching the hot spot, leading to a periodic shedding of the vortex.

The termination region of jets from young stars is very different (see Section 6.3), with at least one of the shocks expected to be radiative. This leads to the formation of a bow-shaped thin shell along the interface between the jet and ambient medium (Blondin *et al.*, 1989). This shell is likely to fragment and may be responsible for time-dependent line fluxes and proper motions of associated knots. In addition, global simulations of dense highly radiative jets reveal that the flow undergoes a transformation before terminating. In both atomic jets (Blondin *et al.*, 1990) and molecular jets (Rosen & Smith, 2004b) (whether light or dense), external shoulders appear well before the termination, which are associated with internal oblique shocks that focus the flow toward the axis. This generates a long protrusion which can efficiently penetrate the ambient medium.

10.2   Jet shapes

A continuous slender jet which does not abruptly widen at any point may be adequately analysed as a one-dimensional flow. This is the simplest description of a steady propagating jet flow in which all the major parameters are functions of just the distance along the channel. This implies that any pressure variations are rapidly communicated across the jet.

In the hydrodynamic twin-exhaust case, the jet cross-sectional area will be determined solely by the changing ambient pressure as the fluid exits the nozzle region and propagates downstream. The ambient pressure acts to confine the jet via the interface, which will consist of an extremely thin turbulent boundary layer. In an inviscid flow description, this is represented by a tangential discontinuity.

Arguments have been presented that the interface will rapidly thicken and the entire jet will inevitably be turbulent due to the high Reynolds number. However, this is only true if the turbulence is able to develop. Within a laminar flow this will require a shear instability to grow under the conditions where expansion acts to reduce the turbulence and advection acts to expel the growing waves. A laminar flow remains laminar for a length termed the potential core length. This length is expected to increase as the Mach number increases and, if it exceeds the pressure scale height, then the turbulence remains suppressed. Although uncertain, the potential core for a light jet may already be about 10 jet radii for a Mach number of $1.5 - 2$ (Bicknell, 1986) and out to about 25 jet radii for a Mach number of 5 (Hardee et al., 1995). If these distances exceed the pressure scale height, then the body of the jet may remain stable.

We first consider the expansion of a non-relativistic isentropic flow. If the pressure does not decrease too fast along the interface, the entire body of the jet will possess the same pressure distribution, $p(z)$, to zeroth order, where z is the distance along the channel from some location $z = z_o$ where $p = p_o$, etc. From the Bernoulli equation, as expressed in Eq. (3.22) ignoring the gravitational term, we find the flow speed, $v(z)$, to be given by

$$v(z)^2 + \frac{2\gamma}{\gamma - 1} \frac{p_o^{1/\gamma}}{\rho_o} p(z)^{1-1/\gamma} = U^2, \tag{10.6}$$

and the cross-sectional area, A, is then given through the constant mass flux, $\rho v A = \rho_o v_o A_o$.

The area can be conveniently written in terms of the conserved total energy flux, $L = \rho v U^2 A/2$, and the stagnation pressure and density, p_s and ρ_s (from a virtual injection location within the cavity with zero speed):

$$A = \frac{2L}{\rho_s U^3} \left(\frac{p}{p_s}\right)^{-1/\gamma} \left[1 - \left(\frac{p}{p_s}\right)^{1-1/\gamma}\right]^{-1/2}. \tag{10.7}$$

Hence, for $p \ll p_s$, the jet speed approaches a terminal speed, U, and the area expands as $A/A_o = (p/p_o)^{-1/\gamma}$.

The appearance of a jet may then be classified by taking the pressure to fall off as a power law, $p/p_o = (z/z_o)^\alpha$. For a circular jet of radius r_s we then obtain

$$\frac{r_s}{r_o} = \left(\frac{z}{z_o}\right)^{-\alpha/2\gamma}. \tag{10.8}$$

In this case the jet will be asymptotically conical if the ambient pressure falls with $\alpha = -2\gamma$. For less steep decreases, the jet boundaries are concave and for steeper fall-offs, the boundaries will be convex. As expected, the flow is divergent (r increases monotonically with z) for all negative values of α.

The relativistic version for the one-dimension flow can be derived from Section 3.6. The constant particle number flux is, from Eq. (3.52), $N\Gamma v A$ where Γ is the Lorentz factor. The constant energy flux is $w\Gamma^2 v c A$, which therefore reduces to the flow constant $w\Gamma/N$.

For an ultra-relativistic equation of state, the enthalpy is $w/4 = p \propto N^{4/3}$. In terms of the stagnation pressure, the area is given by

$$A = \frac{L}{4cp_s} \left(\frac{p}{p_s}\right)^{-1/2} \left[1 - \left(\frac{p}{p_s}\right)^{1/2}\right]^{-1/2}. \tag{10.9}$$

Therefore, we find that, once the fluid has exited the nozzle, the Lorentz factor and the jet circular radius both increase. As the pressure continues to fall, they both approach the same dependence: $\Gamma \propto p^{-1/4}$ and $r \propto p^{-1/4}$ (e.g. Blandford & Rees, 1974). In other words, the Lorentz factor is proportional to the jet radius for as long as the gas remains ultra-relativistic. It increases gradually with only a slow jet widening as the ambient pressure falls.

The ultra-relativistic approximation will break down once the jet has sufficiently expanded. The internal energy per particle, $3p/N$, decreases in proportion to $1/\Gamma$ although the total energy remains constant. Hence, the flow is adiabatically converting the random internal energy into bulk kinetic energy.

At some distance along the expanding jet, the random speeds of particles will become sub-relativistic. If an electron–proton plasma is considered, then the protons will probably become sub-relativistic first and the nature of the flow changes. The bulk Lorentz factor will then not significantly increase, while the jet radius increases slightly faster according to the non-relativistic formula. With initially equal average energies for electrons and protons, the electrons may dominate the pressure and $r \propto p^{-3/8}$. Otherwise, if the protons dominate dynamically, then $r \propto p^{-3/10}$ (Blandford & Rees, 1974).

A hydrodynamic jet as described above will not maintain a circular cross-section unless the jet is conical or the streamlines are diverging. If, as is invariably observed, the jet displays recollimation, there is a slight transverse pressure gradient which pulls material towards the jet axis. This gradient balances the centripetal acceleration of the flow along the curved streamlines. A slight perturbation from a circular cross-section then results in an elliptical cross-section with an increasing eccentricity (Smith & Norman, 1981). In other words, for slowly decreasing ambient pressures with $\alpha > -2\gamma$, a circular cross-section is unstable. At the nozzle, however, the progress to a sheet is slowed and the major axis may flip through $90°$. Gravity also influences the final direction of flattening, but its main effect is to separate the transonic and nozzle locations (Smith, 1994b).

At large distance, the eccentricity becomes large. Therefore, the major axis will always approach towards a constant opening angle, while the minor axis responds rapidly to the ambient pressure changes to yield the correct variation in cross-sectional area, still as described above. Furthermore, the flattening jet is susceptible to pinching and warping which will fragment (possibly bifurcate) the jet before it reaches a thin-sheet configuration.

A magnetic field will contribute towards the force balance in most cases of astrophysical interest. In particular, a toroidal field is likely to be dominant along the extended jet, if not the entire jet. In the proximity of a transonic nozzle, however, the field could be predominantly longitudinal since flux conservation yields $B_p \propto r_j^2$. As the jet subsequently expands, any poloidal component will fall off very fast if there is no transverse shear in the jet speed to convert transverse flux. In the context of a potential jet flow, magnetic flux conservation yields a toroidal field component $B_\phi \propto 1/(R_J v_j)$ (Section 3.3).

This is also a general result for jet models which include rotating hydromagnetic winds provided the jet expands in the extended super-Alfvénic section, and is valid for both non-relativistic (Eichler, 1993) and relativistic (Lyubarsky, 2009) derivations. Rotating magnetised flows are also expected to collimate in the field far from the rigid centrifugal zone (Heyvaerts & Norman, 1989). This is achieved by the tension force associated with the toroidal field, B_ϕ, with the toroidal-to-poloidal field ratio of the order of $B_\phi/B_p \sim r_j/r_A$ once $r_j \gg r_A$. A 'z-pinch' then occurs: the component of the Lorentz force is directed radially inwards. Heyvaerts & Norman (1989) demonstrated that two types of flow structure are possible depending upon the asymptotic behaviour of the total current intensity in the jet. In the limit of a vanishing poloidal current, the field lines are paraboloids that fill space. On the other hand, if the current remains finite, then the field lines take on a cylindrical structure.

An active magnetic field has been included in many basic jet simulations. In particular, a toroidal field is now often justified as a direct consequence of favoured launch mechanisms. In the context of ideal magnetohydrodynamics, Clarke *et al.* (1986) incorporated a dominant toroidal field and demonstrated that a substantial backflowing cocoon was absent. Instead, material collected between the forward and reverse shocks to form a long structure termed a *nose cone* which propagates considerably faster than the hydrodynamic equivalent. The jet, especially the nose section, was found to be 'over-pressured' relative to the ambient medium. Simulations of the equivalent relativistic case also produced nose cones (Leismann *et al.*, 2005). However, the above results were all based on two-dimensional simulations with cylindrical symmetry. In three dimensions, the toroidal field is found to lead to strong current-driven kink instabilities which alter the jet direction (Stone & Hardee, 2000; Mignone *et al.*, 2010). Further work has confirmed that the nose cone is indeed absent in three-dimensional simulations (Cerqueira & de Gouveia Dal Pino, 2001).

Jets originating from collimated magnetocentrifugal winds are certainly more complex in structure even on the larger scales. The magnetic hoop stress arises because the field is wound up by rotation. Along the axis, a poloidal component of magnetic field may remain dominant within a fast light flow. This axial flow may arise directly along open field lines anchored to a star as part of a stellar magnetosphere. It would provide a spinal zone which would not only diverge the surrounding magnetic field lines from the axis, a necessary condition for the centrifugal mechanism to operate, but also enhance the jet stability (Anderson *et al.*, 2006). In simulations under the propeller regime, a stellar outflow dominated by a magnetic field is also envisaged (Ustyugova *et al.*, 2006).

10.3   Jet disruption

A pressure gradient in the surrounding envelope or environment of the source will act as a buoyancy force on a pressure-confined jet, resulting in the acceleration discussed above. However, a steady pressure-confined jet will not be able to keep up with abrupt changes in the atmospheric pressure. The result will be the introduction of non-uniform density and velocity profiles and, more critically, turbulence and shock waves.

To maintain a uniform one-dimensional description would require a high jet sound speed so that signals transmitted by weak pressure waves can rapidly communicate the necessary adjustments: $(dr_j/dz)^2 \ll c_s^2/v_j^2$ in the non-relativistic hydrodynamic case and $(dr_j/dz)^2 \ll c_s^2/\Gamma^2 c^2$ in the ultra-relativistic hydrodynamic case, where c_s is the internal sound speed measured in the co-moving frame. However, detachment occurs once $(dr_j/dz)^2 \geq 1/(\mathcal{M}^2 - 1)$ where $\mathcal{M}$ is the (relativistic proper) Mach number (Lake & Boucher, 1987). Consequently, if the pressure decreases sufficiently fast, confinement cannot be maintained. With sound waves unable to be transmitted far, individual fluid elements tend to move in straight lines. Hence, in this case, the jet will evolve into a free conical expansion with a a transverse speed approximately equal to that associated with the point at which pressure equilibrium across the boundary is lost. More generally, the fast magnetohydrodynamic speed should replace the sound speed in the above formulae.

On the other hand, a free or over-pressured jet may become reconfined if the jet propagates through a zone in which the ambient pressure gradient is shallow. Laing & Bridle (2002a) consider the flaring point in the 3C 31 jets to be one such location where the reconfinement of an initially free jet is achieved through a stationary shock configuration. Furthermore, if the shock causes the flow to transform from supersonic to subsonic flow, through a Mach disc or transverse shock, disruption occurs. Norman *et al.* (1988) demonstrated that such an internal shock is generated if the jet cuts through a shock wave in the external medium. This shock was proposed to be the termination shock from a supersonic wind which blows out from a galactic nucleus, creating a pressure *wall*.

Rather than a pressure jump, Smith (1982) demonstrated that an atmospheric transition with a sharp change in gradient would also disrupt a jet. Extragalactic jets may traverse through an inner zone with a steep, radially decreasing pressure in a cluster cooling flow into a constant pressure intra-cluster medium. In this case, if the pressure in the inner zone falls off somewhat faster than $p \propto z^{-2}$, then the jet degenerates after becoming relatively over-expanded or under-pressured.

The atmosphere itself may not be either steady or uniform. Although collapse speeds are likely to be subsonic, a high-density atmosphere can generate a significant ram pressure on an advancing light jet. Loken *et al.* (1993) explored this scenario, finding that the development of the radio galaxy depends on the jet Mach number. As might be suspected, jets of high Mach number ($M_j > 50$) propagate regardless through the cooling flow. In contrast, low-Mach-number jets ($M_j = 3$) will stagnate as the ram pressure of the atmosphere resists the jet. Interestingly, jets with intermediate Mach numbers are unstable and disrupt within the cooling flow to form amorphous structures. Although these jets overcome the ram pressure, Kelvin–Helmholtz instabilities are found to cause the disruption.

Jets have been fired into a variety of other inhomogeneous ambient media in a wide range of numerical experiments. The collisions with large dense clouds are motivated by strong observational evidence for jet–cloud interactions in radio galaxies (Brodie *et al.*, 1985; van Breugel *et al.*, 1985; Best *et al.*, 1997, see Section 4.3) and protostellar environments (Noriega-Crespo *et al.*, 1996; Reipurth *et al.*, 1996). Wiita *et al.* (2002) demonstrated that powerful light jets will eventually disperse a cloud, while weak jets can be effectively stopped by reasonably massive clouds. Jet instabilities can be enhanced, especially for lower-Mach-number jets.

The interaction with an off-axis cloud was simulated by de Gouveia Dal Pino (1999) and Raga *et al.* (2002a) in the two contrasting contexts. The initial interaction may appear as a reflection feature, especially with a light jet and a more direct impact. This soon develops into a bent jet structure The deflected beam initially cuts a curved trajectory due to the directed ram pressure along the contact discontinuity. The deflection angle is not fixed (Zhang *et al.*, 1999) but depends on the collision details and varies with time. The deflection angle can well exceed 90° and could, in theory, even take a hair-pin track if the deflection angle could be held constant (Smith, 1984). However, the jet gradually straightens with time as the jet penetrates the cloud in both cases. Once the jet has penetrated the cloud, the deflected structure decays and the jet resumes its original path. The transient nature of a jet–cloud interaction was also pointed out by De Young (1991). During the interaction itself, complex internal structure is generated including numerous knots and distorted shells which have been interpreted in terms of Herbig–Haro objects (Raga & Canto, 1995).

It is also clear that small-scale inhomogeneities which lie in the jet path influence the jet propagation and produce complex structures. The degree of degeneration depends on the filling factor of the clumps (Saxton *et al.*, 2005; Yirak *et al.*, 2008), with jet disruption occuring primarily as a result of Kelvin–Helmholtz instabilities driven by turbulence in the radio cocoon rather than through direct jet–clump interactions. For ballistic jets, Moraghan *et al.* (2008) found that thick dense sheets are easily penetrated by a dense jet in a similar fashion to a bullet penetrating a target. In contrast to fixed obstructions, Jeyakumar (2009) simulated moving clouds which can enter the jet and temporarily block it until either exiting on the other side or eroding material into the jet flow.

10.4 Jet flares and knots

The extreme blazar TeV flares (see Section 5.5) can be produced by strong shocks formed far out in the region where the energy of a Poynting jet is eventually converted somehow into matter and radiation. This is at distances exceeding 10^4 Schwarzschild radii (Sikora *et al.*, 2009). Such shocks may be formed internally as a result of instabilities which generate flow regions with different speeds. Or, as argued by Sikora *et al.* (2008), a standing reconfinement shock can better explain the 2005 outburst in the quasar 3C454.3 in addition to the HST-1 feature of M 87 (Section 5.1.3). Moreover, the latter knot is at a putative distance of 10 parsecs from the black hole, in a location where the high-energy radiation could originate as inverse Compton scattering of infrared photons emitted by extended hot dust. A standing shock will dissipate jet energy more efficiently than an internal shock.

In the case of M 87, the argument for a stationary reconfinement shock is more explicitly expressed by Stawarz *et al.* (2006). The jet itself is considered to originate as a high-pressure flow undergoing a free expansion. The pressure in the conical jet expands according to a law given by Eq. (10.8): $p_j(z) \propto z^{2\gamma}$ where the power-law index could be as steep as $-10/3$ for a cold jet. In contrast, the ambient pressure possesses a much shallower pressure gradient in the same region. Therefore, such a jet would become reconfined at some point with a converging/diverging reconfinement/reflected shock pattern (see Fig. 10.4) dominating the volume of the jet.

The reconfinement model, originally discussed by Falle & Wilson (1985), was extended by Tsinganos & Bogovalov (2002) to include the interaction between a strongly magnetised relativistic inner jet and a non-relativistic collimating disc wind. They demonstrated that a converging–diverging shock pattern in the relativistic jet can be expected.

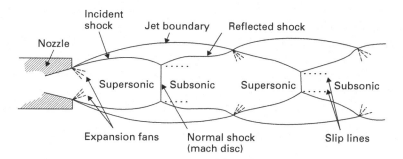

Fig. 10.4 The convergent–divergent flow pattern associated with a pressure-imbalanced supersonic jet entering an ambient medium. Extracted from Norman *et al.* (1982).

An alternative mechanism to form flares within the Poynting jet itself invokes magnetic reconnection to produce *mini-jets*, which are highly relativistic flows superimposed on and transverse to the main jet. The propagation is transverse to the main axis, which provides a solution to the M 87 phenomenon where the jet cannot be strongly beamed (Giannios *et al.*, 2010), as discussed in Section 5.1.3. If two emitting regions appear in the reconnecting process associated with a magnetic island, then, as required, the relative importance of the inverse Compton process is increased (Nalewajko *et al.*, 2011).

Another interpretation of the knots in the M 87 jet was presented by Blandford and Koenigl (1979). It was proposed that they are the sites of dense blobs of gas lying within a jet. The dense blobs are coincident with either supernova remnants or interstellar clouds, which thus act as obstacles. Jet kinetic energy is dissipated behind a bow shock associated with each obstacle, where particle acceleration and magnetic field amplification take place to produce the emission across the spectrum. The obstacles evolve through the incident momentum flux, possibly fragmenting into a string of knots. Material is also accelerated in the jet direction, appearing as knots until their speed approaches that of the jet and the knots fade.

The proper motions of radio knots in quasar and microquasar jets, especially the super-luminal motions (Section 5.2), support models incorporating travelling shock waves. This was first suggested for the M 87 optical knots by Rees (1978), who suggested that velocity irregularities would steepen into internal shock waves. The general properties are consistent with the propagation of relativistic magnetohydrodynamic shock fronts (Eichler & Smith, 1983) within helically twisted magnetic structures (Nakamura *et al.*, 2010). However, the origin of the shock waves remains obscure. There is considerable evidence from black hole X-ray transients such as GRS 1915+105 that the appearance of a knot is triggered by the disappearance of the inner annuli of the accretion disc (Mirabel *et al.*, 1998).

The knots from jets stemming from young stars are also triggered by pulsations associated with the developing source (see Section 6.1.4). The optical knots are successfully modelled as weak shock waves within a fast jet (Raga *et al.*, 1990) and can be shown to trace the ejection history (Raga *el al.*, 2002b). Many numerical simulations have run on this theme. In particular, Stone and Norman (1993) provided a detailed interpretation of the dynamics. Each smooth, large-amplitude pulse rapidly develops into a twin-shock configuration as it propagates down the jet. An upstream shock propagates more slowly than the jet and decelerates high-speed jet material as it catches up with the pulse, while a downstream shock sweeps up lower-speed material ahead of it. Sandwiched between the shocks is a high-pressure layer which gradually thickens as well as ejecting material laterally. Consequently, the pulses decay in amplitude.

In the molecular case associated with protostars, the cooling in the dense knots is sufficient for the entire jet to develop into a decaying saw-tooth structure (Smith *et al.*, 1997). Three-dimensional simulations of pulses within a uniform jet display the classical pattern of transverse shock structures internal to the jet (Fig. 10.5) while either a velocity shear or spray across the jet results in internal bow shocks and a highly complex terminal working surface (Völker *et al.*, 1999).

Fig. 10.5 The simulation of a pulsed molecular jet of speed $100\,\mathrm{km\,s^{-1}}$ entering a molecular medium of lower density. The scale unit is 10^{16} cm, time is 604 years and the greyscale represents the logarithm of the H_2 density per cm^3. The pulse amplitude is $30\,\mathrm{km\,s^{-1}}$ and period is 50 years. Taken from Suttner *et al.* (1997).

10.5   Instability

Although some jets appear perfectly straight over limited stretches, most jets at least display kinks and wiggles in the sky plane. These may be associated with sinusoidal or helical motions of the underlying bulk flow in three dimensions. The wiggles may be amplified downstream and lead to wide plumes or complete disruption of the interior as well as the surface layers. Jets also invariably display a pattern of changes in width which could be the result of pinches and knots attributed to axisymmetric structure. Most interpretations for the development of these structures fall into four broad classes (Nakamura *et al.*, 2007), as follows.

1.   Magnetohydrodynamic or fluid dynamic instabilities such as the Kelvin–Helmholtz instability (KHI)
2.   Current-driven instability (CDI)
3.   Pressure-driven instability (PDI)
4.   Precession of the jet ejection axis.

Which instabilities operate will, of course, depend on the nature of the jet, which can be considered as either kinetic flux-dominated (KFD) or Poynting flux-dominated (PFD) (Nakamura *et al.*, 2007). It is well known that a cylindrical plasma column with a helical magnetic configuration is subject to the $m = 0$ (sausage) mode of the PDI, the $m = 1$ (kink or screw) mode of the CDI, and other higher-order modes, where m is the azimuthal mode number.

The Kelvin–Helmholtz instability occurs at the interface between two media in relative motion. In the jet context, a supersonic flow is assumed to be pressure-confined by a stationary thermal medium. However, global simulations have shown that the jet vacates a cavity as it ploughs through the ambient medium (Norman *et al.*, 1982). The cavity is filled with the overspill from the jet resulting from the geometry of the impact zone where the jet material is shocked. The cocoon is particularly wide when the jet is of high Mach number and light: the ambient density far exceeds the jet density. In this case, the shocked jet material is raised to a high pressure and must expand laterally. Hence, the jet actually interacts directly with a cocoon of hot ejected gas rather than a quiescent ambient medium.

The Kelvin–Helmholtz instability was first analysed with extragalactic jets in mind along with the development of the purely adiabatic hydrodynamic model (Hardee, 1981, 1987). Although the observed jets are not disrupted out to hundreds of kiloparsecs, the instability was still thought to be responsible for generating knot structure and wiggles as well as mixing the two media (Ferrari *et al.*, 1978). In these light jets, mixing could lead to significant mass transfer from the ambient medium into the jet (entrainment) and momentum deposition into the ambient medium. In the jet, knots and wiggles can both arise through the non-linear development of axisymetric and helical-type modes. These modes are saturated through shocks appearing as either weak internal or lateral shocks which dissipate a small fraction of the entire bulk flow energy.

Jet simulations can be split into three domains according to the initial conditions. One assumption is that the jet flows through a prescribed ambient medium across the entire grid, with no backflow or cocoon. These *cross-grid simulations* of *equilibrium jets* could conceivably occur if a jet either sheds its cocoon as it evolves or the jet enters a prepared vacated channel formed by an earlier outflow phase. If the grid is sufficiently long, the spatial growth of perturbations can be followed. The second type considers a fixed portion of an *infinite periodic jet*. This allows one to follow the temporal evolution of a wave perturbation assumed to be initially imposed along the entire jet. However, it should be noted that the growth length of the instability is sensitive to the choice of spatial or temporal domain for the analysis (Birkinshaw, 1984).

The alternative *global simulations* of *propagating jets* start with an undisturbed ambient medium occupying the entire simulated region. The jet is then switched on or ramped up. In this case, the jet interacts with its cocoon even after the jet has crossed the grid. Depending on the chosen conditions, the jet and cocoon pressures can far exceed that of the ambient medium since there is insufficient time for sound waves to cross the entire configuration.

The wave modes can be split into two types, surface and body modes. Surface modes are standing unstable modes which decay exponentially with distance away from the interface. They are present under all flow conditions but are damped by an axial magnetic field or relativistic motion. The shortest wavelengths possess the highest growth rates, leading to a turbulent interface of saturated waves if the boundary is sharp. Otherwise, the minimum wavelength will depend on the thickness of the sheared boundary. The body modes are reflected overstable travelling perturbations inherent only to supersonic jet flows. These appear under all conditions and are found to dominate jet structures.

The instability has been analysed in detail analytically and in two dimensions through cross-grid simulations of slab rather than cylindrical jets (Hardee & Norman, 1988; Norman & Hardee, 1988). According to the linearised dispersion relations, cylindrical jets are subject to pinch, helical and fluting modes, each of which has fundamental and reflection solutions. As a result of perturbations to the body of the jet, the rapid growth of resonant pinch modes, which reflect back and forth from the jet boundaries, leads to internal oblique crossing shocks. These, however, are not necessarily disruptive. On the other hand, the helical modes with the azimuthal wave mode number $m = \pm 1$ are the most dangerous to the jet integrity (Ferrari, 1998). The jet is disrupted within $\sim 10 \times M_j r_j$ due to the growth of these surface disturbances. The manner in which this is achieved depends on the driving frequency and the wave resonant frequency. It also depends on the opening angle of the jet with expansion able to suppress the growth of surface waves and subsequent turbulence (Rosen & Hardee, 2000)

In temporal simulations, reflected modes are found to dominate, while ordinary surface modes are limited to the slowest-growing long-wavelength modes (Bodo *et al.*, 1994). Axisymmetric simulations demonstrate three phases: an initial linear growth phase leading to the formation of concical shock waves, followed by the strengthening of the shocks and associated modulation of the jet structure, and finally a mixing and disruption phase (Bodo *et al.*, 1994).

In the presence of radiative cooling, the KHI growth rates and wavelengths can be damped or amplified in comparison to the adiabatic case (Hardee & Stone, 1997). Thermal instability appears to work beside the KHI to promote the instability in the presence of a flat cooling curve (as a function of temperature) but to suppress the instability if the cooling is a steep positive function of temperature. Reflection shocks can appear amplified through the shock compression along with a complex interaction between body and surface waves, as displayed in Fig. 10.6. In cooling jets, the shocks can thus produce dense knots and filaments of cooling gas within the jet. Moreover, ripples in the jet surface drive spur-like oblique shocks into the ambient medium which can set in motion a large amount of ambient gas to low speeds. This was illustrated in three-dimensional simulations by Xu *et al.* (2000) which suggest that the KHI may be a means to produce the wide low-speed bipolar outflows associated with protostars (Downes & Ray, 1998).

The KHI analyses have incorporated special relativistic effects (Turland & Scheuer, 1976; Hardee, 2000), super-Alfvénic or superfast magnetosonic flows (Hardee *et al.*, 1997) and their combination (Hardee, 2007). The magnetic field partially stabilises the flow in comparison with purely hydrodynamic jets of the same Mach number (e.g. Gardiner *et al.*, 2000). These works also indicate that the presence of a jet sheath or a trans- or super-Alfvénic relativistic jet spine can provide a stable backbone (Mizuno *et al.*, 2007). However, mainly parallel magnetic fields have been considered in these spine–sheath configurations, and jet rotation, toroidal and helical fields add to the complexity.

The current-driven instability (CDI) is driven by an electrical current flowing parallel to the jet axis. It can occur even in a zero-pressure, force-free plasma. PFD jets are particularly prone to destruction because they contain a strong axial current and a corresponding highly wound helical field, which results in a dynamically dominant Lorentz force.

Several analytic studies of the CDI have been performed. A linear analysis of non-relativistic force-free jets which are thermally confined by an external medium was presented by Appl & Camenzind (1992). It was found that current-carrying jets are

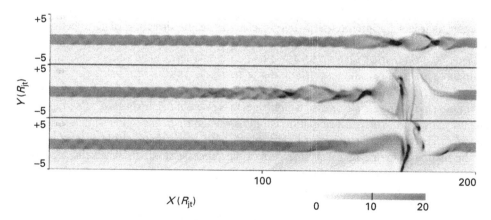

Fig. 10.6 Greyscale images of the density for simulations of a slab jet which is adiabatic (upper panel), and with two alternative cooling functions which reprent atomic cooling. The jet is perturbed with a frequency $\omega r_j/v_j = 1.0$ at the nozzle. The jet is overdense by a factor of 10 and the internal Mach number is 50, or Mach 5 relative to the ambient sound speed. Figure extracted from Stone *et al.* (1997). Reproduced by permission of the AAS.

substantially more stable than their current-free counterparts. Furthermore, the growth rate of the CDI is smaller than that of the KHI for super-magnetosonic flows, being comparable only for quite small fast-magnetosonic numbers. Appl (1996) concludes that the integrity of super-magnetosonic collimated jets is not endangered by the CDI.

More general current-carrying magnetic field configurations have been analysed and simulated with a diversity of results according to the chosen set-up. While both the CDI and KHI produce helical twists through kink instabilities, the CDI produces a very different structure. The current density is redistributed within the inner part of the jet on the Alfvén crossing timescale (Lery *et al.*, 2000). The CDI-driven structure then propagates at the jet speed and is prominent internal to a spine–sheath interface and moves at the jet speed.

A jet with a strong toroidal magnetic field is highly prone to disruption by the current-driven kink or screw instability, as reviewed by McKinney & Blandford (2009). The criterion for the potentially most disruptive $m = 1$ mode is given by the Kruskal–Shafranov (KS) instability criterion (e.g. Li, 2000),

$$-\frac{B_\phi}{B_p} > \frac{2\pi R}{r}, \tag{10.10}$$

for cylindrical force-free equilibria where R is the cylindrical radius, r is the poloidal extent such that $R = r \sin \theta$. This indicates that a forming jet is liable to be unstable very soon after the Alfvén point where the toroidal field becomes dominant. However, the work of McKinney & Blandford (2009) demonstrates that there are various effective stabilising effects related to the rotation, shear and expansion which weaken the above criterion. These effects then permit mildly relativistic jet speeds to be attained on large scales.

10.6   Changing direction

10.6.1   Precession and wiggling

We have encountered a huge number of jets of all types which display curved structure that can be interpreted as due to changes in the direction of jet ejection. The descriptions cover all possibilities from wiggling to wide-angle precession. In particular, Galactic X-ray binaries such as SS 433 (Section 7.5.4) have been observed with highly predictable precessing radio jets (Hjellming & Johnston, 1981).

The launch axis may change in time for various reasons. The jet axis may correspond to the rotation axis of an accretion disc which precesses or wobbles. In the case of extragalactic jets, the cause of the disc re-orientation could be related to an encounter with a secondary galactic core or the existence of a binary black hole. Supermassive black hole binaries are the natural outcome of galaxy mergers. Tidal precession occurs when the secondary black hole is in a close non-coplanar orbit with the primary accretion disc (Romero *et al.*, 2000) and can produce precession periods of order of years. In contrast, the Lense–Thirring dragging of inertial frames through misalignment between the disc and a Kerr black hole (Begelman *et al.*, 1980) generates precession with a period of at least a thousand years.

For stellar jets, the jet source may well be part of a binary system. In this case, the jet would take on the source's orbital motion around the binary companion in addition to a precessional motion of the warped accretion disc (Raga *et al.*, 2009). The orbital motion would produce a mirror-symmetric jet pair on small scales, as found in HH 111

(Section 6.2.1. On the other hand, the disc axis precession period is an order of magnitude longer (Terquem *et al.*, 1999), thus generating a point-symmetric jet pair on large scales. Another general cause of disc wobbling is irradiation from the central source which causes the disc to warp (Pringle, 1996).

The first substantial simulations of astrophysical jets with precession were undertaken by Cox *et al.* (1991) once three-dimensional hydrodynamics was achievable. Since then, a large number of simulations have been performed but generally still lack resolution. Of the results, the clearest occur for heavy jets which behave ballistically. As expected, low precession produces an open spiral that appears as a curving ribbon on the sky in many shock-tracing emission lines (e.g. Smith & Rosen, 2005). If pulses are superimposed, then the helical path degenerates into a chain of bow shocks. In contrast, simulations involving a fast precession resemble that expected from a conical jet with a ring-like working surface (Fig. 10.7). As the ring spreads, it loses coherency and fragments into a complex of curved arcs.

10.6.2 Bending

A fast cross-wind or headwind can strongly deflect an exposed jet. In the case of radio galaxies, the driving source may be moving at a speed of a few thousand kilometres per second through a quite dense intracluster medium. This realisation prompted Begelman *et al.* (1979) to develop a model for the narrow-angled tail radio galaxies (see Section 4.1.3) in which the twin jets are hit by a side-wind. Once exposed or naked, it is the direct ram

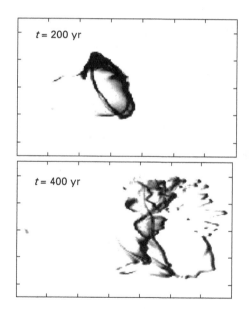

Fig. 10.7 Greyscale images of the molecular hydrogen emission from a fast-processing jet. Early on (top panel), a coherent ring is produced. Later (lower panel), segments of the dense ring propagate independently and become unstable. H_2 emission from the 1–0 S(1) transition is displayed; the jet speed is $100 \, \mathrm{km \, s^{-1}}$ and precession (half-)angle is $20°$. The code employed is ZEUS-3D on a $355 \times 230 \times 230$ grid with zone size 2×10^{14} cm. Taken from Rosen & Smith (2004a).

pressure of the intergalactic medium which provides the bending force. The flow through a side-wind has been simulated many times (e.g. Williams & Gull, 1984; Balsara & Norman, 1992), demonstrating that internal oblique shocks and the Kelvin–Helmholtz instability are critical in the bending and disruption process.

An alternative model considered the pressure gradients within gas bound to the galaxy as the bending agent (Jones & Owen, 1979). The gradients are enhanced by the external ram pressure due to the galactic motion. This model assumes that a large elliptical galaxy can retain a quite extended distribution of dense gas, perhaps out to 10 kpc. The induced gradients are then responsible for bending the jets. Both mechanisms were tested in the simulations of Soker *et al.* (1988), who concluded that induced pressure gradients are much less effective in bending jets.

Some jets associated with young stars and protostars are also suspected to be subject to supersonic side-winds possibly generating C-type or mirror symmetries. The wind can be the result of the peculiar motion of the jet source or a large-scale motion of the ambient interstellar cloud. This problem has been tackled analytically by Canto & Raga (1995) and numerically by Lim and Raga (1998), who show that even ballistic jets can be effectively bent through large angles.

A transverse force will be exerted on a pressure-confined jet as it propagates through any non-spherical pressure distribution. Even the hydrostatic gas in a galaxy will sustain pressure gradients in the interstellar medium which are still expected to bend a jet unless it is ejected directly down the pressure gradient itself. Such gradual systematic bending of pressure-confined jets was first analysed by Smith & Norman (1981). It was found that the trajectories were significantly altered only for low Mach numbers with the maximum bending, under the most favourable conditions, being inversely proportional to the Mach number. In the relativistic case, this angle is the inverse of the jet Lorentz factor in radians.

10.7   Summary

Jet propagation, termination and associated instabilities have been extensively theoretically investigated, especially in the matter- or kinetic energy-dominated regime. At present, physical assumptions, as well as chosen initial and boundary conditions, dominate the diverse set of results obtained. As observations improve, the chosen states become better motivated and, the hope is, we will finally link up the models of launching with those of propagation.

11

The astrophysical jet

From the first eight chapters, we conclude that the subject of astrophysical jets has seen observational advances over the last twenty years that take it almost beyond recognition in breadth and depth. The penultimate two chapters have shown how the major theories from the 1980s have been developed and adapted but not dismissed. Yet, there are long-standing debates which have not found a resolution and new issues which stretch our understanding.

The challenging themes can be listed under the following headings. We still need to gather evidence for:

- The process of formation and launch
- The collimation and acceleration
- The propagation, structure and stability
- The contents, both near and far from the driving source.

Here, rather than providing a long list of questions, we focus on some specific hot issues relevant to the general subject: the composition, regulation, feedback and unification.

11.1   Composition

The physical parameters, such as the energy, speed, mass flux and abundances of jet material, can often be constrained after identifying the radiation process. However, invisible components, whose presence has not yet been settled, may be flowing out adjacent to the radiating material or mixed in on the particle scale. For example, in the observed warm atomic and cold molecular jets from young stars and protostars, there may be more outflowing mass in the form of cold neutral atomic gas (Nisini et al., 2005). In addition, dust internal to the jet avoids direct detection (Dionatos et al., 2009).

The content of extragalactic jets has proved difficult to fully specify, with no spectral features to unambiguously identify the material. Assuming matter dominates, blazar jets have been hypothesised to be either electron–proton (Celotti & Fabian, 1993) or electron–positron in content. From radio observations of synchrotron emission we know that relativistic electrons and magnetic field are present in the jets. However, under the assumption of charge neutrality, there must also be positively charged particles. The nature of these particles has remained obscure since they are not directly measurable through observations.

Protons are favoured on arguments related to the bulk kinetic energy supplied to the extended sources and on constraints related to pair annihilation (Celotti & Fabian, 1993).

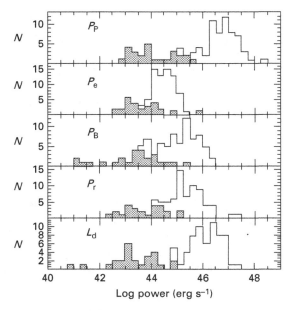

Fig. 11.1 Estimations for the components which carry the power in a sample of blazar jets. The top panel displays the bulk kinetic power transmitted by cold protons (P_p, assuming one proton per emitting electron); middle panel: emitting electrons (P_e, random and bulk); third panel: Poynting flux (P_B) assuming minimum energy arguments; fourth panel: radiation (P_r). The lowest panel shows the distribution of the luminosities (L_d) of the associated accretion discs. The shaded areas correspond to sources with only upper limits on L_d. Extracted from Ghisellini *et al.* (2010).

In terms of the bulk flow energy required to produce the large-scale radio lobes and X-ray cavities, an electron–proton jet is more plausible (Celotti & Ghisellini, 2008). As shown in Fig. 11.1, one proton per emitting electron will solve the power budget and leave enough power to be carried to the radio lobes (Celotti & Ghisellini, 2008).

Positrons are favoured for the M 87 jet where higher densities are indicated by the core self-absorption properties, whereas the low kinetic flux required of 10^{43} erg s^{-1} (per jet) to supply the radio lobes (Reynolds *et al.*, 1996) can be supplied without protons. In terms of the minimum energy required in the jets themselves, an electron–positron plasma would require the least amount of energy. The signal at 511 keV from electron–positron two-photon annihilation has proved elusive, with the broadening and shifting due to the high speeds adding to the difficulty (Marscher *et al.*, 2007). Thus, from the blazar in isolation, a leptonic model is sufficient with electrons or electron–positron pairs capable of producing all the synchrotron and inverse Compton emission.

However, the cavities surrounding radio galaxies place severe constraints on the kinetic energy which must be supplied, suggesting that cold protons may dominate the energy transfer even though the number of electron–positron pairs may far exceed the number of protons (De Young, 2006). Nevertheless, the best available data are still consistent with an electron–positron scenario (Dunn *et al.*, 2006).

What do the launch models predict? A jet driven off from an accretion disc is likely to be composed of normal electron–proton material. In the Poynting-flux scenario, high-energy photons are generated which interact with the electromagnetic field to produce electron–positron material. Celotti and Ghisellini (2008) demonstrated that the Poynting flux cannot still dominate in the region generating inverse Compton emission since this severely limits the magnetic field energy. To complicate the issue, the contents may be contaminated well after launch through mass loading processes such as stellar winds (Hubbard & Blackman, 2006).

An alternative theory is based on the so-called *proton blazar* or hadronic model, which invokes the presence of highly relativistic protons. These protons emit directly via the synchrotron process or undergo proton–proton or proton–photon interactions before triggering electron–positron pair cascades (Mannheim, 1993). This model has some difficulties associated with pair reprocessing.

The composition of a jet associated with a gamma-ray burst remains controversial but must include photons, magnetic field and matter (e.g. Zhang *et al.*, 2011). Photons are advected with the matter and magnetic field until the optical depth to Compton scattering falls below unity. Further along the jet, photons are still generated behind shock waves and reconnection zones, and escape freely.

The magnetic component of a GRB jet is considered as a Poynting flux with an induced electric field in the observer's frame. The matter can be split into baryonic and leptonic components. The baryonic component will clearly dominate the mass unless the proton-to-electron number ratio is very low. Protons will dominate over heavy ions since the latter will not survive the relativistic conditions. In addition, free neutrons are expected initially, coupled with the protons through strong interactions. These neutrons then decouple and decay with a co-moving life time of 900 s. However, the nature of the jets is still tentative, as discussed by Fan (2009).

11.2 Regulation

The association between accretion discs and jets is seen in many, but not all, systems. It may thus be that jets are necessary to permit disc accretion. In the inflow–outflow connection, some fraction of the accreting material reaches a growing central source while the rest is ejected along with all the excess angular momentum. Hence, jets can be considered to be an intimate part of the system forming a star or black hole rather than a by-product (Ferreira, 1997). In fact, almost the entire inflow of mass could be ejected, so directly regulating the growth of the central object. The strong interplay between accretion and ejection is postulated in the scenario which generates the so-called magnetised accretion-ejection structures. In this scenario, accretion is not achieved via a standard accretion disc in which angular momentum is transported radially outward through the disc by turbulent viscosity, but through the extraction of spin employing magnetocentrifugal processes. This must still be mediated by a diffusion process, either ambipolar diffusion or turbulent diffusion, to enable matter to traverse the magnetic surfaces.

In the X-wind and spinning black hole scenarios, the jet extraction of angular momentum from the central object could be considered non-essential. These scenarios provide consistent interpretations for the radio-quiet quasars which contain black holes that appear to have their mass regulated despite the absence of jets (e.g. Di Matteo *et al.*, 2005). One solution involves the regulation of radio-quiet objects by neighbouring radio-loud objects in an early dense protocluster environment (Rawlings & Jarvis, 2004). Furthermore, recent data provide evidence for a minimum accretion rate being necessary to produce a given jet power (Fernandes *et al.*, 2011), with some jet powers exceeding that available from release through accretion (Punsly, 2007, 2011; McNamara *et al.*, 2011). If confirmed, the spinning black hole model would be strongly favoured

There is also a growing argument for protostellar and young stellar jets to be driven from the interaction zone of a disc with a rotating magnetosphere rather than through the disc alone. Simulations have shown that the system is self-regulating (Matt *et al.*, 2002), with the large-scale magnetic field structure dominating the dynamics and evolving to a configuration that removes disc angular momentum as well as determining the mass outflow rate. Further simulations demonstrate that the stellar magnetosphere may well play a crucial role (Fendt, 2009; Romanova *et al.*, 2009).

11.3   Feedback

Contrary to intuition, light jets appear to have more of an impact on their surroundings than heavy jets. Despite the low momentum flux, light jets are capable of producing cluster-scale cavities given sufficient time. Furthermore, the momentum flux can be efficiently transferred into the ambient medium. More importantly, this flux can be converted via turbulence into heat energy, and so support the intracluster gas as required for structure formation in conventional cosmological models (e.g. De Young, 2010). Hence, the interpretation of X-ray cavities (Section 4.2.1) shows that AGN feedback regulates galactic star formation and suppresses cooling of the hot halos.

On the other hand, heavy molecular jets do not create extensive cavities. They have the ability to penetrate the ambient gas without transferring a high fraction of their momentum flux. It follows from Eq. (10.2) that the efficiency of transfer of jet momentum to the ambient medium is $1/(1 + \sqrt{\eta})^2$, which is low if the ratio of jet to ambient density, η, is high. Therefore, heavy jets are not expected to provide strong feedback into the immediate surroundings of the molecular cloud. However, outflow feedback from forming stars in a cluster-forming cloud is thought to help regulate the star formation efficiency but not halt the protocluster collapse (Maury *et al.*, 2009).

11.4   Unification

A grand unification scheme for all AGNs will require at least three key factors to be specified: orientation, intrinsic luminosity and Eddington ratio (Antonucci, 1993; Urry & Padovani, 1995). The first one is certain: the viewing angle. As illustrated in Fig. 11.2, radio quasars are related to FR II radio galaxies such that the axis of the jet is close to the line of sight. Similarly, the BL Lac objects are related to the FR I radio galaxies, together being a class of intrinsically lower luminosity. The jet orientation is related to the axis of an extended dusty molecular torus which obscures the direct view of the continuum and broad-line emission region. Note, however, that obscured quasars also exist since the cores of galaxies are likely to stockpile gas and dust at all orientations during their violent episodes of formation.

The above objects all harbour a similar central engine, differing in type according to the black hole mass and accretion rate. In addition, the spin of the black hole may determine

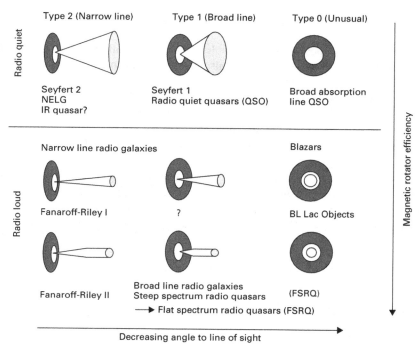

Fig. 11.2 Standard classification of AGN sources following Urry and Padovani (1995). The horizontal axis represents the inclination of the source axis with the line of sight. The vertical axis represents a parameter that is linked to the efficiency of the underlying magnetic rotator. Figure adapted from Meliani *et al.* (2010).

whether an AGN is radio-loud or radio-quiet. Further differences may be related to the galactic environment in which the relativistic jets propagate.

Evolutionary unification schemes for radio galaxies and protostellar outflows are still being considered. Compact symmetric objects are either very young radio galaxies (Owsianik & Conway, 1998) or frustrated radio galaxies (Carvalho, 1998) depending on their ability to bore their way through the interstellar medium. Giant radio galaxies and relic radio sources may represent the final throes. For young stars the sequence may be the reverse, with the observable sections of the jets getting shorter as a star ages (Stanke, 2003; Davis *et al.*, 2009).

11.5 The future

The above issues suggest that there are many more discoveries to come. Orientation affects our view of not only AGNs, but microquasar and gamma-ray burst systems also. The orientation also influences the observed properties of young stars, with jet-cleared tunnels allowing direct views into the engines of some protostars.

Rotation is probably a feature of all jets yet is extremely difficult to measure conclusively in terms of line emission or jet shape. Cometary jets may be subject to a tornado effect as sublimating ice spins up as it approaches the eye of a vent. However, even for fast-rotating comets, this is a minor effect.

The magnetic field poses a similar problem. Its origin, configuration and amplification all remain unclear if not mysterious. It is also not clear what conditions would produce and sustain the ordered field surfaces often proposed.

Finally, jets and accretion discs are closely related in X-ray binaries and microquasars where the ejection of a radio component is often synchronous with the disappearance of plasma in the inner disc (a low in the high-energy flux). For protostars and young stars of low and moderate mass, jets are found exclusively in objects which harbour accretion discs. Moreover, as a young star evolves, the jets become less prominent as accreting material is depleted (the infrared excess decreases). For active galactic nuclei, the emission from radio jets and lobes is correlated with the disc emission (inferred from optical line emission). Therefore, the long-held view connecting jets to discs now takes on considerable weight. Nevertheless, as shown in Chapter 8, jets are also quite easy to configure without discs.

References

Abdo A. A., Ackermann M., Ajello M., Axelsson, M., Baldini L., Ballet J., Barbiellini G., Bastieri D., Baughman B. M., Bechtol K. *et al.*, 2010, *Nature*, **463**, 919

Abraham Z., Romero G. E., 1999, *A&A*, **344**, 61

Abraham Z., Carrara E. A., Zensus J. A., Unwin S. C., 1996, *A&AS*, **115**, 543

Acciari V. A., Aliu E., Arlen T., Aune T., Beilicke M., Benbow W., Boltuch D., Bradbury S. M., Massaro F., 2010, *ApJ*, **716**, 819

Agudo I., Bach U., Krichbaum T. P., Marscher A. P., Gonidakis I., Diamond P. J., Perucho M., Alef W., Graham D. A., Witzel A., Zensus J. A., Bremer M., Acosta-Pulido J. A., Barrena R., 2007, *A&A*, **476**, L17

Aharonian F., Akhperjanian A. G., Bazer-Bachi A. R., Behera B., Beilicke M., Benbow W., Berge D., Bernlöhr K., Boisson C., Bolz O., Borrel V., Boutelier T., Braun I., Brion E., Brown A. M., Zdziarski A. A., 2007, *ApJ*, **664**, L71

Albert J., Aliu E., Anderhub H., Antoranz P., Armada A., Baixeras C., Barrio J. A., Bartko H., Bastieri D., Becker J. K., Bednarek W., Berger K., 2007, *ApJ*, **669**, 862

Albert J., Aliu E., Anderhub H., Antonelli L. A., Antoranz P., Backes M., Baixeras C., Barrio J. A., Bartko H., Bastieri D., Becker J. K., Bednarek W., Zapatero J., 2008, *ApJ*, **685**, L23

Anderson J. M., Li Z., Krasnopolsky R., Blandford R. D., 2005, *ApJ*, **630**, 945

Anderson J. M., Li Z.-Y., Krasnopolsky R., Blandford R. D., 2006, *ApJ*, **653**, L33

Andre P., Ward-Thompson D., Barsony M., 1993, *ApJ*, **406**, 122

Angelini L., White N. E., 2003, *ApJ*, **586**, L71

Anglada G., 1996, in A. R. Taylor & J. M. Paredes eds., *Radio Emission from the Stars and the Sun*, Vol. 93 of Astronomical Society of the Pacific Conference Series, Radio Jets in Young Stellar Objects. pp. 3–7

Anglada G., Rodriguez L. F., Canto J., Estalella R., Torrelles J. M., 1992, *ApJ*, **395**, 494

Anglada G., López R., Estalella R., Masegosa J., Riera A., Raga A. C., 2007, *AJ*, **133**, 2799

Antonucci R., 1993, *ARA&A*, **31**, 473

Antoniucci S., Nisini B., Giannini T., Lorenzetti D., 2008, *A&A*, **479**, 503

Appl S., 1996, *A&A*, **314**, 995

Appl S., Camenzind M., 1992, *A&A*, **256**, 354

Asai A., Shibata K., Hara H., Nitta N. V., 2008, *ApJ*, **673**, 1188

Aschenbach B., Brinkmann W., 1975, *A&A*, **41**, 147

Ashbourn J. M. A., Woods L. C., 2005, *Physics Scripta*, 71, 123

Aspin C., Schwarz H. E., Smith M. G., Corradi R. L. M., Mountain C. M., Wright G. S., Ramsay S. K., Robertson D., Beard S. M., Pickup D. A., Geballe T. R., Bridger A., Laird D., Montgomery D., Glendinning R., Pentland G., Griffin J. L., Aycock J., 1993, *A&A*, **278**, 255

Baade W., 1956, *ApJ*, **123**, 550

Baade W., Minkowski R., 1954, *ApJ*, **119**, 215

Bacciotti F., Eislöffel J., 1999, *A&A*, **342**, 717

Bacciotti F., Chiuderi C., Oliva E., 1995, *A&A*, **296**, 185

Bacciotti F., Eislöffel J., Ray T. P., 1999, *A&A*, **350**, 917

Bacciotti F., Mundt R., Ray T. P., Eislöffel J., Solf J., Camezind M., 2000, *ApJ*, **537**, L49

Bacciotti F., Ray T. P., Mundt R., Eislöffel J., Solf J., 2002, *ApJ*, **576**, 222

Bachiller R., 1996, *ARA&A*, **34**, 111

Bahcall J. N., Kirhakos S., Schneider D. P., Davis R. J., Muxlow T. W. B., Garrington S. T., Conway R. G., Unwin S. C., 1995, *ApJ*, **452**, L91

Bai J. M., Lee M. G., 2003, *ApJ*, **585**, L113
Bain H. M., Fletcher L., 2009, *A&A*, **508**, 1443
Balbus S. A., Hawley J. F., 1991, *ApJ*, **376**, 214
Bally J., Lada C. J., 1983, *ApJ*, **265**, 824
Bally J., Reipurth B., 2003, *AJ*, **126**, 893
Bally J., Devine D., Reipurth B., 1996, *ApJ*, **473**, L49
Bally J., Feigelson E., Reipurth B., 2003, *ApJ*, **584**, 843
Balsara D. S., Norman M. L., 1992, *ApJ*, **393**, 631
Banerjee R., Klessen R. S., Fendt C., 2007, *ApJ*, **668**, 1028
Bar-Nun A., Pálsson F., Björnsson H., 2008, Icarus, 197, 164
Barthel P. D., 1989, *ApJ*, **336**, 606
Baum S. A., O'Dea C. P., Dallacassa D., de Bruyn A. G., Pedlar A., 1993, *ApJ*, **419**, 553
Baum S. A., O'Dea C. P., de Koff S., Sparks W., Hayes J. J. E., Livio M., Golombek D., 1996, *ApJ*, **465**, L5
Becker C. M., Remillard R. A., Rappaport S. A., McClintock J. E., 1998, *ApJ*, **506**, 880
Beckers J. M., 1968, *Solar Physics*, **3**, 367
Begelman M. C., Li Z., 1994, *ApJ*, **426**, 269
Begelman M. C., Rees M. J., Blandford R. D., 1979, *Nature*, **279**, 770
Begelman M. C., Blandford R. D., Rees M. J., 1980, *Nature*, **287**, 307
Begelman M. C., Blandford R. D., Rees M. J., 1984, *Reviews of Modern Physics*, 56, 255
Begelman M. C., Fabian A. C., Rees M. J., 2008, *MNRAS*, **384**, L19
Belloni T., Mendez M., King A. R., van der Klis M., van Paradijs J., 1997, *ApJ*, **479**, L145
Belton M. J. S., Melosh J., 2009, *Icarus*, **200**, 280
Benford G., 1978, *MNRAS*, **183**, 29
Berger E., Kulkarni S. R., Pooley G., Frail D. A., McIntyre V., Wark R. M., Sari R., Soderberg A. M., Fox D. W., Yost S., Price P. A., 2003, *Nature*, **426**, 154
Bessel F. W., 1836, *Astronomische Nachrichten*, **13**, 185
Best P. N., Longair M. S., Rottgering H. J. A., 1997, *MNRAS*, **286**, 785
Beuther H., Schilke P., Gueth F., 2004, *ApJ*, **608**, 330
Bicknell G. V., 1984, *ApJ*, **286**, 68
Bicknell G. V., 1986, *ApJ*, **300**, 591
Biretta J. A., Sparks W. B., Macchetto F., 1999, *ApJ*, **520**, 621
Biretta J. A., Junor W., Livio M., 2002, *New Astronomy Review*, **46**, 239
Birkinshaw M., 1984, *MNRAS*, **208**, 887
Blackman E. G., Perna R., 2004, *ApJ*, **601**, L71
Blandford R. D., Königl A., 1979a, *Astrophysical Letterrs*, **20**, 15
Blandford R. D., Königl A., 1979b, *ApJ*, **232**, 34
Blandford R. D., McKee C. F., 1976, *Physics of Fluids*, **19**, 1130
Blandford R. D., Payne D. G., 1982, *MNRAS*, **199**, 883
Blandford R. D., Rees M. J., 1974, *MNRAS*, **169**, 395
Blandford R. D., Znajek R. L., 1977, *MNRAS*, **179**, 433
Blondin J. M., Königl A., Fryxell B. A., 1989, *ApJ*, **337**, L37
Blondin J. M., Fryxell B. A., Königl A., 1990, *ApJ*, **360**, 370
Blundell K. M., Bowler M. G., 2004, *ApJ*, **616**, L159
Blundell K. M., Rawlings S., 2001, *ApJ*, **562**, L5
Blundell K. M., Bowler M. G., Schmidtobreick L., 2008, *ApJ*, **678**, L47
Bodo G., Massaglia S., Ferrari A., Trussoni E., 1994, *A&A*, **283**, 655
Boehringer H., Voges W., Fabian A. C., Edge A. C., Neumann D. M., 1993, *MNRAS*, **264**, L25
Boice D. C., Britt D. T., Nelson R. M., Sandel B. R., Soderblom L. A., Thomas N., Yelle R. V., 2002a, in *Lunar and Planetary Institute Science Conference Abstracts*, Vol. 33 of Lunar and Planetary Inst. Technical Report, The Near-Nucleus Environment of 19P/Borrelly During the Deep Space One Encounter. pp. 1810
Boice D. C., Soderblom L. A., Britt D. T., Brown R. H., Sandel B. R., Yelle R. V., Buratti B. J., Hicks M. D., Nelson R. M., Rayman M. D., Oberst J., Thomas N., 2002b, *Earth Moon and Planets*, **89**, 301
Borkowski K. J., Blondin J. M., Harrington J. P., 1997, *ApJ*, **482**, L97
Böttcher M., Reimer A., Marscher A. P., 2009, *ApJ*, **703**, 1168
Bridle A. H., 1984, *AJ*, **89**, 979
Bridle A. H., Perley R. A., 1984, *ARA&A*, **22**, 319

Brocksopp C., Sokoloski J. L., Kaiser C., Richards A. M., Muxlow T. W. B., Seymour N., 2004, *MNRAS*, **347**, 430

Brodie J. P., Bowyer S., McCarthy P., 1985, *ApJ*, **293**, L59

Brooks K. J., Garay G., Voronkov M., Rodríguez L. F., 2007, *ApJ*, **669**, 459

Buehrke T., Mundt R., Ray T. P., 1988, *A&A*, **200**, 99

Bujarrabal V., Alcolea J., Sánchez Contreras C., Sahai R., 2002, *A&A*, **389**, 271

Burns J. O., Loken C., Roettiger K., Rizza E., Bryan G., Norman M. L., Gómez P., Owen F. N., 2002, *New Astronomy Review*, **46**, 135

Burrows C. J., Stapelfeldt K. R., Watson A. M., Krist J. E., Ballester G. E., Clarke J. T., Crisp D., Gallagher III J. S., Griffiths R. E., Hester J. J., Hoessel J. G., Holtzman J. A., Mould J. R., Scowen P. A., Trauger J. T., Westphal J. A., 1996, *ApJ*, **473**, 437

Burton C. E., 1882, *MNRAS*, **42**, 422

Cabrit S., Pety J., Pesenti N., Dougados C., 2006, *A&A*, **452**, 897

Cai M. J., Shang H., Lin H., Shu F. H., 2008, *ApJ*, **672**, 489

Calvet N., Muzerolle J., Briceño C., Hernández J., Hartmann L., Saucedo J. L., Gordon K. D., 2004, *AJ*, **128**, 1294

Camenzind M., 1987, *A&A*, **184**, 341

Canfield R. C., Reardon K. P., Leka K. D., Shibata K., Yokoyama T., Shimojo M., 1996, *ApJ*, **464**, 1016

Canto J., Raga A. C., 1995, *MNRAS*, **277**, 1120

Canto J., Rodriguez L. F., 1980, *ApJ*, **239**, 982

Canto J., Tenorio-Tagle G., Rozyczka M., 1988, *A&A*, **192**, 287

Capetti A., Fanti R., Parma P., 1995, *A&A*, **300**, 643

Caratti o Garatti A., Giannini T., Nisini B., Lorenzetti D., 2006, *A&A*, **449**, 1077

Carilli C. L., Perley R. A., Dreher J. H., 1988, *ApJ*, **334**, L73

Carvalho J. C., 1998, *A&A*, **329**, 845

Celotti A., Fabian A. C., 1993, *MNRAS*, **264**, 228

Celotti A., Ghisellini G., 2008, *MNRAS*, **385**, 283

Cernicharo J., Reipurth B., 1996, *ApJ*, **460**, L57

Cerqueira A. H., de Gouveia Dal Pino E. M., 2001, *ApJ*, **550**, L91

Chatterjee R., Jorstad S. G., Marscher A. P., Oh H., McHardy I. M., Aller M. F., Aller H. D., Balonek T. J., Miller H. R., Ryle W. T., Tosti G., Kurtanidze O., Nikolashvili M., Larionov V. M., Hagen-Thorn V. A., 2008, *ApJ*, **689**, 79

Chernin L. M., Masson C. R., 1995, *ApJ*, **443**, 181

Cheung C. C., Harris D. E., Stawarz Ł., 2007, *ApJ*, **663**, L65

Chiu H. H., 1973, *Physics of Fluids*, **16**, 825

Christiansen W. A., Scott J. S., Vestrand W. T., 1978, *ApJ*, **223**, 13

Cirtain J. W., Golub L., Lundquist L., van Ballegooijen A., Savcheva A., Shimojo M., DeLuca E., Tsuneta S., Sakao T., Reeves K., Weber M., Kano R., Narukage N., Shibasaki K., 2007, *Science*, **318**, 1580

Clarke D. A., Bridle A. H., Burns J. O., Perley R. A., Norman M. L., 1992, *ApJ*, **385**, 173

Clarke D. A., Norman M. L., Burns J. O., 1986, *ApJ*, **311**, L63

Claussen M. J., Marvel K. B., Wootten A., Wilking B. A., 1998, *ApJ*, **507**, L79

Codella C., Cabrit S., Gueth F., Cesaroni R., Bacciotti F., Lefloch B., McCaughrean M. J., 2007, *A&A*, **462**, L53

Coffey D., Bacciotti F., Woitas J., Ray T. P., Eislöffel J., 2004, *ApJ*, **604**, 758

Coffey D., Bacciotti F., Podio L., 2008, *ApJ*, **689**, 1112

Cohen M. H., Moffet A. T., Romney J. D., Schilizzi R. T., Seielstad G. A., Kellermann K. I., Purcell G. H., Shaffer D. B., Pauliny-Toth I. I. K., Preuss E., 1976, *ApJ*, **206**, L1

Colwell J. E., Jakosky B. M., Sandor B. J., Stern S. A., 1990, *Icarus*, **85**, 205

Contopoulos J., 1994, *ApJ*, **432**, 508

Contopoulos J., Lovelace R. V. E., 1994, *ApJ*, **429**, 139

Conway R. G., Davis R. J., 1994, *A&A*, **284**, 724

Coppin K. E. K., Davis C. J., Micono M., 1998, *MNRAS*, **301**, L10

Corradi R. L. M., Munari U., Livio M., Mampaso A., Gonçalves D. R., Schwarz H. E., 2001, *ApJ*, **560**, 912

Corradi R. L. M., Perinotto M., Villaver E., Mampaso A., Gonçalves D. R., 1999, *ApJ*, **523**, 721

Correia S., Zinnecker H., Ridgway S. T., McCaughrean M. J., 2009, *A&A*, **505**, 673

Costa E., Frontera F., Heise J., Feroci M., in't Zand J., Fiore F., Cinti M. N., Dal Fiume D., Nicastro L., Orlandini M., Palazzi E., Rapisarda M., 1997, *Nature*, **387**, 783

Cowley A. P., Schmidtke P. C., Crampton D., Hutchings J. B., 1998, *ApJ*, **504**, 854

Cox C. I., Gull S. F., Scheuer P. A. G., 1991, *MNRAS*, **252**, 558

Cox P., Lucas R., Huggins P. J., Forveille T., Bachiller R., Guilloteau S., Maillard J. P., Omont A., 2000, *A&A*, **353**, L25

Crampton D., Hutchings J. B., Cowley A. P., Schmidtke P. C., McGrath T. K., O'Donoghue D., Harrop-Allin M. K., 1996, *ApJ*, **456**, 320

Crocker M. M., Davis R. J., Eyres S. P. S., Bode M. F., Taylor A. R., Skopal A., Kenny H. T., 2001, *MNRAS*, **326**, 781

Crocker M. M., Davis R. J., Spencer R. E., Eyres S. P. S., Bode M. F., Skopal A., 2002, *MNRAS*, **335**, 1100

Curtis H. D., 1918, *Publications of Lick Observatory*, **13**, 9

Daigne F., Mochkovitch R., 2002, *MNRAS*, **336**, 1271

Daly R. A., Marscher A. P., 1988, *ApJ*, **334**, 539

Davis C. J., Smith M. D., Eislöffel J., Davies J. K., 1999, *MNRAS*, **308**, 539

Davis C. J., Kumar M. S. N., Sandell G., Froebrich D., Smith M. D., Currie M. J., 2007, *MNRAS*, **374**, 29

Davis C. J., Froebrich D., Stanke T., Megeath S. T., Kumar M. S. N., Adamson A., Eislöffel J., Gredel R., Khanzadyan T., Lucas P., Smith M. D., Varricatt W. P., 2009, *A&A*, **496**, 153

Davis C. J., Gell R., Khanzadyan T., Smith M. D., Jenness T., 2010, *A&A*, **511**, A24

de Gouveia dal Pino E. M., 1999, *ApJ*, **526**, 862

de Gouveia dal Pino E. M., Benz W., 1993, *ApJ*, **410**, **686**

De Pontieu B., Hansteen V. H., Rouppe van der Voort L., van Noort M., Carlsson M., 2007a, *ApJ*, **655**, 624

De Pontieu B., McIntosh S., Hansteen V. H., Carlsson M., Schrijver C. J., Tarbell T. D., Title A. M., Shine R. A., Suematsu Y., Tsuneta S., Katsukawa Y., Ichimoto K., Shimizu T., Nagata S., 2007b, *PASJ*, **59**, 655

De Pontieu B., McIntosh S. W., Carlsson M., Hansteen V. H., Tarbell T. D., Boerner P., Martinez-Sykora J., Schrijver C. J., Title A. M., 2011, *Science*, **331**, 55

De Villiers J., Hawley J. F., Krolik J. H., 2003, *ApJ*, **599**, 1238

de Vries W. H., Becker R. H., White R. L., 2006, *AJ*, **131**, 666

De Young D. S., 1991, *ApJ*, **371**, 69

De Young D. S., 2006, *ApJ*, **648**, 200

De Young D. S., 2010, *ApJ*, **710**, 743

Demoulin M., Burbidge G. R., 1968, *ApJ*, **154**, 3

Dennis T. J., Cunningham A. J., Frank A., Balick B., Blackman E. G., Mitran S., 2008, *ApJ*, **679**, 1327

Desmurs J., Codella C., Santiago-García J., Tafalla M., Bachiller R., 2009, *A&A*, **498**, 753

Dhawan V., Mirabel I. F., Rodríguez L. F., 2000, *ApJ*, **543**, 373

Di Matteo T., Springel V., Hernquist L., 2005, *Nature*, **433**, 604

Dionatos O., Nisini B., Garcia Lopez R., Giannini T., Davis C. J., Smith M. D., Ray T. P., DeLuca M., 2009, *ApJ*, **692**, 1

Dopita M. A., Evans I., Schwartz R. D., 1982, *ApJ*, **263**, L73

Double G. P., Baring M. G., Jones F. C., Ellison D. C., 2004, *ApJ*, **600**, 485

Dougados C., Cabrit S., Lavalley C., Ménard F., 2000, *A&A*, **357**, L61

Downes T. P., Ray T. P., 1998, *A&A*, **331**, 1130

Draine B. T., McKee C. F., 1993, *ARA&A*, **31**, 373

Draine B. T., Roberge W. G., Dalgarno A., 1983, *ApJ*, **264**, 485

Dreher J. W., Feigelson E. D., 1984, *Nature*, **308**, 43

Dunlop J. S., McLure R. J., Kukula M. J., Baum S. A., O'Dea C. P., Hughes D. H., 2003, *MNRAS*, **340**, 1095

Dunn R. J. H., Fabian A. C., Celotti A., 2006, *MNRAS*, **372**, 1741

Eichler D., 1993, *ApJ*, **419**, 111

Eichler D., Smith M., 1983, *Nature*, **303**, 779

Eilek J. A., Owen F. N., 2002, *ApJ*, **567**, 202

Eisloeffel J., Mundt R., 1992, *A&A*, **263**, 292

Eisloeffel J., Mundt R., 1994, *A&A*, **284**, 530

Eislöffel J., Mundt R., 1998, *AJ*, **115**, 1554

Elitzur M., ed. 1992, *Astronomical Masers*, Vol. 170 of Astrophysics and Space Science Library

Fabian A. C., Sanders J. S., Ettori S., Taylor G. B., Allen S. W., Crawford C. S., Iwasawa K., Johnstone R. M., Ogle P. M., 2000, *MNRAS*, **318**, L65

Falle S. A. E. G., Wilson M. J., 1985, *MNRAS*, **216**, 79

Fan J., Huang Y., He T., Yang J. H., Hua T. X., Liu Y., Wang Y. X., 2009, *PASJ*, **61**, 639

Fan Y.-Z., 2009, *MNRAS*, **397**, 1539

Fanaroff B. L., Riley J. M., 1974, *MNRAS*, **167**, 31P

Farnham T. L., 2009, *Planetary and Space Science*, **57**, 1192

Farnham T. L., Samarasinha N. H., Mueller B. E. A., Knight M. M., 2007a, *AJ*, **133**, 2001

Farnham T. L., Wellnitz D. D., Hampton D. L., Li J., Sunshine J. M., Groussin O., McFadden L. A., Crockett C. J., A'Hearn M. F., Belton M. J. S., Schultz P., Lisse C. M., 2007b, *Icarus*, **187**, 26

Feibelman W. A., 1985, *AJ*, **90**, 2550

Fender R., Spencer R., Tzioumis T., Wu K., van der Klis M., van Paradijs J., Johnston H., 1998, *ApJ*, **506**, L121

Fendt C., 2006, *ApJ*, **651**, 272

Fendt C., 2009, *ApJ*, **692**, 346

Fendt C., Zinnecker H., 1998, *A&A*, **334**, 750

Fermi LAT Collaboration Abdo A. A., *et al.* 2009, *Science*, **326**, 1512

Fermi LAT Collaboration Abdo A. A., *et al.* 2010, *Science*, **328**, 725

Fernandes C. A. C., Jarvis M. J., Rawlings S., Martínez-Sansigre A., Hatziminaoglou E., Lacy M., Page M. J., Stevens J. A., Vardoulaki E., 2011, *MNRAS*, **411**, 1909

Ferrari A., 1998, *ARA&A*, **36**, 539

Ferrari A., Trussoni E., Zaninetti L., 1978, *A&A*, **64**, 43

Ferreira J., 1997, *A&A*, **319**, 340

Ferreira J., Pelletier G., 1995, *A&A*, **295**, 807

Ferrín I., 2010, *Planetary and Space Science*, **58**, 1868

Flower D. R., Pineau des Forets G., Field D., May P. W., 1996, *MNRAS*, **280**, 447

Fomalont E. B., Ebneter K. A., van Breugel W. J. M., Ekers R. D., 1989, *ApJ*, **346**, L17

Fomalont E. B., Geldzahler B. J., Bradshaw C. F., 2001, *ApJ*, **558**, 283

Fortes A. D., 2007, *Icarus*, **191**, 743

Frail D. A., Kulkarni S. R., Sari R., Djorgovski S. G., Bloom J. S., Galama T. J., Reichart D. E., Berger E., Harrison F. A., Price P. A., Yost S. A., Diercks A., Goodrich R. W., Chaffee F., 2001, *ApJ*, **562**, L55

Frank A., 1999, *New Astronomy Review*, **43**, 31

Frank A., Mellema G., 1996, *ApJ*, **472**, 684

Fukue J., 2000, *PASJ*, **52**, 613

Fukue J., Shibata K., Okada R., 1991, *PASJ*, **43**, 131

Gaensler B. M., Arons J., Kaspi V. M., Pivovaroff M. J., Kawai N., Tamura K., 2002, *ApJ*, **569**, 878

Galloway D. K., Sokoloski J. L., 2004, *ApJ*, **613**, L61

Garay G., Brooks K. J., Mardones D., Norris R. P., 2003, *ApJ*, **587**, 739

García-Lario P., Riera A., Manchado A., 1999, *ApJ*, **526**, 854

Garcia Lopez R., Nisini B., Giannini T., Eislöffel J., Bacciotti F., Podio L., 2008, *A&A*, **487**, 1019

Gardiner T. A., Frank A., Jones T. W., Ryu D., 2000, *ApJ*, **530**, 834

Garofalo D., Evans D. A., Sambruna R. M., 2010, *MNRAS*, **406**, 975

Georganopoulos M., Kazanas D., 2003a, *ApJ*, **594**, L27

Georganopoulos M., Kazanas D., 2003b, *ApJ*, **589**, L5

Ghisellini G., Padovani P., Celotti A., Maraschi L., 1993, *ApJ*, **407**, 65

Ghisellini G., Maraschi L., Dondi L., 1996, *A&AS*, **120**, C503

Ghisellini G., Celotti A., Costamante L., 2002, *A&A*, **386**, 833

Ghisellini G., Tavecchio F., Chiaberge M., 2005, *A&A*, **432**, 401

Ghisellini G., Tavecchio F., Foschini L., Ghirlanda G., Maraschi L., Celotti A., 2010, *MNRAS*, **402**, 497

Ghosh P., Abramowicz M. A., 1997, *MNRAS*, **292**, 887

Giannini T., McCoey C., Caratti o Garatti A., Nisini B., Lorenzetti D., Flower D. R., 2004, *A&A*, **419**, 999

Giannios D., Uzdensky D. A., Begelman M. C., 2010, *MNRAS*, **402**, 1649

Gieseking F., Becker I., Solf J., 1985, *ApJ*, **295**, L17

Gilbert G. M., Riley J. M., Hardcastle M. J., Croston J. H., Pooley G. G., Alexander P., 2004, *MNRAS*, **351**, 845

Ginzburg V. L., Syrovatskii S. I., 1965, *ARA&A*, **3**, 297

Ginzburg V. L., Syrovatskii S. I., 1969, *ARA&A*, **7**, 375

Giuliani A., D'Ammando F., Vercellone S., Vittorini V., Chen A. W., Donnarumma I., Pacciani L., Pucella G., Trois A., Bulgarelli A., Longo F., Tavani M., Tosti G., Impiombato D., Argan A., 2009, *A&A*, **494**, 509

Gizani N. A. B., Leahy J. P., 2003, *MNRAS*, **342**, 399

Gombosi T. I., Horanyi M., 1986, *ApJ*, **311**, 491

Gomez P. L., Pinkney J., Burns J. O., Wang Q., Owen F. N., Voges W., 1997, *ApJ*, **474**, 580

Goodson A. P., Winglee R. M., Boehm K., 1997, *ApJ*, **489**, 199

Gopal-Krishna, Wiita P. J., 2000, *A&A*, **363**, 507

Gopal-Krishna, Wiita P. J., 2001, *A&A*, **373**, 100

Gopal-Krishna, Dhurde S., Wiita P.J., 2004, *ApJ*, **615**, L81

Grupe D., Gronwall C., Wang X., Roming P. W. A., Cummings J., Zhang B., Mészáros P., Trigo M. D., O'Brien P. T., Page K. L., Beardmore A., Sakamoto T., Gehrels N., 2007, *ApJ*, **662**, 443

Güdel M., Skinner S. L., Audard M., Briggs K. R., Cabrit S., 2008, *A&A*, **478**, 797

Guerrero M. A., Miranda L. F., Riera A., Velázquez P. F., Olguín L., Vázquez R., Chu Y., Raga A., Benítez G., 2008, *ApJ*, **683**, 272

Gueth F., Guilloteau S., 1999, *A&A*, **343**, 571

Guilloteau S., Dutrey A., Pety J., Gueth F., 2008, *A&A*, **478**, L31

Gurman J. B., Thompson B. J., Newmark J. A., Deforest C. E., 1998, in R. A. Donahue & J. A. Bookbinder eds, *Cool Stars, Stellar Systems, and the Sun*, Vol. 154 of Astronomical Society of the Pacific Conference Series, New Images of the Solar Corona. p 329

Hanasz M., Sol H., 1996, *A&A*, **315**, 355

Hansen C. J., Esposito L. W., Stewart A. I. F., Meinke B., Wallis B., Colwell J. E., Hendrix A. R., Larsen K., Pryor W., Tian F., 2008, *Nature*, **456**, 477

Hardcastle M. J., 1998, *MNRAS*, **298**, 569

Hardcastle M. J., 1999, *A&A*, **349**, 381

Hardcastle M. J., Sakelliou I., 2004, *MNRAS*, **349**, 560

Hardcastle M. J., Alexander P., Pooley G. G., Riley J. M., 1999, *MNRAS*, **304**, 135

Hardcastle M. J., Birkinshaw M., Worrall D. M., 2001, *MNRAS*, **326**, 1499

Hardee P. E., 1981, *ApJ*, **250**, L9

Hardee P. E., 1987, *ApJ*, **313**, 607

Hardee P. E., 2000, *ApJ*, **533**, 176

Hardee P. E., 2007, *ApJ*, **664**, 26

Hardee P. E., Norman M. L., 1988, *ApJ*, **334**, 70

Hardee P. E., Stone J. M., 1997, *ApJ*, **483**, 121

Hardee P. E., Clarke D. A., Howell D. A., 1995, *ApJ*, **441**, 644

Hardee P. E., Clarke D. A., Rosen A., 1997, *ApJ*, **485**, 533

Harker D. E., Wooden D. H., Woodward C. E., Lisse C. M., 2002, *ApJ*, **580**, 579

Haro G., 1952, *ApJ*, **115**, 572

Harris D. E., Krawczynski H., 2006, *ARA&A*, **44**, 463

Harris D. E., Cheung C. C., Stawarz Ł., Biretta J. A., Perlman E. S., 2009, *ApJ*, **699**, 305

Harrison T. E., Bornak J., Rupen M. P., Howell S. B., 2010, *ApJ*, **710**, 325

Hartigan P., 1989, *ApJ*, **339**, 987

Hartigan P., Hillenbrand L., 2009, *ApJ*, **705**, 1388

Hartigan P., Morse J., 2007, *ApJ*, **660**, 426

Hartigan P., Raymond J., Meaburn J., 1990, *ApJ*, **362**, 624

Hartigan P., Morse J. A., Reipurth B., Heathcote S., Bally J., 2001, *ApJ*, **559**, L157

Hartigan P., Heathcote S., Morse J. A., Reipurth B., Bally J., 2005, *AJ*, **130**, 2197

Hartmann L., Calvet N., Gullbring E., D'Alessio P., 1998, *ApJ*, **495**, 385

He J., Marsch E., Curdt W., Tian H., Tu C., Xia L., Kamio S., 2010, *A&A*, **519**, A49

Hedman M. M., Nicholson P. D., Showalter M. R., Brown R. H., Buratti B. J., Clark R. N., 2009, *ApJ*, **693**, 1749

Heggland L., De Pontieu B., Hansteen V. H., 2009, *ApJ*, **702**, 1

Helfand D. J., Gotthelf E. V., Halpern J. P., 2001, *ApJ*, **556**, 380

Henney W. J., O'Dell C. R., Meaburn J., Garrington S. T., Lopez J. A., 2002, *ApJ*, **566**, 315

Henri G., Saugé L., 2006, *ApJ*, **640**, 185

Herbig G. H., 1951, *ApJ*, **113**, 697

Herbig G. H., 1980, *IAU Circular*, **3535**, 2

Hevelius J., 1682, *Acta Eruditorum.* Dec., p.389

Heyvaerts J., Norman C., 1989, *ApJ*, **347**, 1055

Heyvaerts J., Priest E. R., 1984, *A&A*, **137**, 63

Heywood I., Blundell K. M., Rawlings S., 2007, *MNRAS*, **381**, 1093

Hirano N., Liu S., Shang H., Ho P. T. P., Huang H., Kuan Y., McCaughrean M. J., Zhang Q., 2006, *ApJ*, **636**, L141

Hirose S., Uchida Y., Shibata K., Matsumoto R., 1997, *PASJ*, **49**, 193

Hirth G. A., Mundt R., Solf J., Ray T. P., 1994, *ApJ*, **427**, L99

Hirth G. A., Mundt R., Solf J., 1997, *A&AS*, **126**, 437

Hjellming R. M., Johnston K. J., 1981, *ApJ*, **246**, L141

Hjorth J., Sollerman J., Møller P., Fynbo J. P. U., Woosley S. E., Kouveliotou C., Tanvir N. R., Greiner J., Andersen M. I., Castro-Tirado A. J., Castro Cerón J. M., Fruchter A. S., Vreeswijk P. M., Watson D., Wijers R. A. M. J., 2003, *Nature*, **423**, 847

Hogbom J. A., 1979, *A&AS*, **36**, 173

Hollis J. M., Michalitsianos A. G., 1993, *ApJ*, **411**, 235

Homan D. C., Ojha R., Wardle J. F. C., Roberts D. H., Aller M. F., Aller H. D., Hughes P. A., 2001, *ApJ*, **549**, 840

Homan D. C., Ojha R., Wardle J. F. C., Roberts D. H., Aller M. F., Aller H. D., Hughes P. A., 2002, *ApJ*, **568**, 99

Hovatta T., Valtaoja E., Tornikoski M., Lähteenmäki A., 2009, *A&A*, **494**, 527

Hoyle F., 1966, *Nature*, **209**, 751

Hubbard A., Blackman E. G., 2006, *MNRAS*, **371**, 1717

Huebner W. F., Buhl D., Snyder L. E., 1974, *Icarus*, **23**, 580

Huebner W. F., Boice D. C., Reitsema H. J., Delamere W. A., Whipple F. L., 1988, *Icarus*, **76**, 78

Ichimaru S., 1977, *ApJ*, **214**, 840

Iijima T., Esenoglu H. H., 2003, *A&A*, **404**, 997

Ip W., Mendis D. A., 1977, *Ap&SS*, **46**, 109

Jetha N. N., Hardcastle M. J., Sakelliou I., 2006, *MNRAS*, **368**, 609

Jeyakumar S., 2009, *Astronomische Nachrichten*, **330**, 287

Jiang Y. C., Chen H. D., Li K. J., Shen Y. D., Yang L. H., 2007, *A&A*, **469**, 331

Jones P. A., 1990, *Proceedings of the Astronomical Society of Australia*, **8**, 254

Jones T. W., Owen F. N., 1979, *ApJ*, **234**, 818

Jorstad S. G., Marscher A. P., Lister M. L., Stirling A. M., Cawthorne T. V., Gómez J., Gear W. K., 2004, *AJ*, **127**, 3115

Jorstad S. G., Marscher A. P., Lister M. L., Stirling A. M., Cawthorne T. V., Gear W. K., Gómez J. L., Stevens J. A., Smith P. S., Forster J. R., Robson E. I., 2005, *AJ*, **130**, 1418

Kaaret P., Corbel S., Tomsick J. A., Lazendic J., Tzioumis A. K., Butt Y., Wijnands R., 2006, *ApJ*, **641**, 410

Kargaltsev O. Y., Pavlov G. G., Teter M. A., Sanwal D., 2003, *New AR*, **47**, 487

Karovska M., Gaetz T. J., Carilli C. L., Hack W., Raymond J. C., Lee N. P., 2010, *ApJ*, **710**, L132

Karpen J. T., Antiochos S. K., Devore C. R., 1995, *ApJ*, **450**, 422

Kato M., Hachisu I., 2003, *ApJ*, **587**, L39

Kato Y., Mineshige S., Shibata K., 2004, *ApJ*, **605**, 307

Katz-Stone D. M., Rudnick L., Butenhoff C., O'Donoghue A. A., 1999, *ApJ*, **516**, 716

Keller H. U., Delamere W. A., Reitsema H. J., Huebner W. F., Schmidt H. U., 1987, *A&A*, **187**, 807

Keller H. U., Knollenberg J., Markiewicz W. J., 1994, *Planetary and Space Science*, **42**, 367

Kellermann K. I., Pauliny-Toth I. I. K., 1969, *ApJ*, **155**, L71

Kellermann K. I., Kovalev Y. Y., Lister M. L., Homan D. C., Kadler M., Cohen M. H., Ros E., Zensus J. A., Vermeulen R. C., Aller M. F., Aller H. D., 2007, *Ap&SS*, **311**, 231

Kellogg E., Anderson C., Korreck K., DePasquale J., Nichols J., Sokoloski J. L., Krauss M., Pedelty J., 2007, *ApJ*, **664**, 1079

Khanzadyan T., Smith M. D., Davis C. J., Stanke T., 2004, *A&A*, **418**, 163

Kharb P., O'Dea C. P., Baum S. A., Daly R. A., Mory M. P., Donahue M., Guerra E. J., 2008, *ApJS*, **174**, 74

Kieffer H. H., Christensen P. R., Titus T. N., 2006, *Nature*, **442**, 793

Kieffer S. W., Lu X., Bethke C. M., Spencer J. R., Marshak S., Navrotsky A., 2006, *Science*, **314**, 1764

Kieffer S. W., Lu X., McFarquhar G., Wohletz K. H., 2009, *Icarus*, **203**, 238

King A. R., Pringle J. E., 2009, *MNRAS*, **397**, L51

Kitamura Y., 1986, *Icarus*, **66**, 241

Klebesadel R. W., Strong I. B., Olson R. A., 1973, *ApJ*, **182**, L85

Königl A., 1980, *Physics of Fluids*, **23**, 1083

Königl A., 1982, *ApJ*, **261**, 115

Königl A., Choudhuri A. R., 1985a, *ApJ*, **289**, 173

Königl A., Choudhuri A. R., 1985b, *ApJ*, **289**, 188

Körding E., Rupen M., Knigge C., Fender R., Dhawan V., Templeton M., Muxlow T., 2008, *Science*, **320**, 1318

Kormendy J., Richstone D., 1995, *ARA&A*, **33**, 581

Kovalev Y. Y., Lister M. L., Homan D. C., Kellermann K. I., 2007, *ApJ*, **668**, L27

Koza J., Rutten R. J., Vourlidas A., 2009, *A&A*, **499**, 917

Kraft R. P., Forman W. R., Jones C., Murray S. S., Hardcastle M. J., Worrall D. M., 2002, *ApJ*, **569**, 54

Krause M., 2003, *A&A*, **398**, 113

Krautter A., Henriksen R. N., Lake K., 1983, *ApJ*, **269**, 81

Kudoh T., Shibata K., 1997, *ApJ*, **474**, 362
Kudoh T., Shibata K., 1999, *ApJ*, **514**, 493
Kusunose M., Takahara F., 2008, *ApJ*, **682**, 784
Labita M., Treves A., Falomo R., Uslenghi M., 2006, *MNRAS*, **373**, 551
Laing R. A., Bridle A. H., 2002a, *MNRAS*, **336**, 1161
Laing R. A., Bridle A. H., 2002b, *MNRAS*, **336**, 328
Laing R. A., Parma P., de Ruiter H. R., Fanti R., 1999, *MNRAS*, **306**, 513
Laing R. A., Canvin J. R., Bridle A. H., 2006, *Astronomische Nachrichten*, **327**, 523
Lake K., Boucher P., 1987, *ApJ*, **314**, 507
Lanza A., Miller J. C., Motta S., 1985, *Physics of Fluids*, **28**, 97
Lavalley-Fouquet C., Cabrit S., Dougados C., 2000, *A&A*, **356**, L41
Lazzati D., Morsony B. J., Begelman M. C., 2010, *ApJ*, **717**, 239
Leahy J. P., Perley R. A., 1991, *AJ*, **102**, 537
Leahy J. P., Williams A. G., 1984, *MNRAS*, **210**, 929
Leahy J. P., Black A. R. S., Dennett-Thorpe J., Hardcastle M. J., Komissarov S., Perley R. A., Riley J. M., Scheuer P. A. G., 1997, *MNRAS*, **291**, 20
Ledlow M. J., Owen F. N., Yun M. S., Hill J. M., 2001, *ApJ*, **552**, 120
Lee C., Ho P. T. P., Hirano N., Beuther H., Bourke T. L., Shang H., Zhang Q., 2007, *ApJ*, **659**, 499
Lee C., Ho P. T. P., Bourke T. L., Hirano N., Shang H., Zhang Q., 2008, *ApJ*, **685**, 1026
Lee C., Hirano N., Palau A., Ho P. T. P., Bourke T. L., Zhang Q., Shang H., 2009, *ApJ*, **699**, 1584
Lee C., Hasegawa T. I., Hirano N., Palau A., Shang H., Ho P. T. P., Zhang Q., 2010, *ApJ*, **713**, 731
Lefloch B., Cernicharo J., Reipurth B., Pardo J. R., Neri R., 2007, *ApJ*, **658**, 498
Leismann T., Antón L., Aloy M. A., Müller E., Martí J. M., Miralles J. A., Ibáñez J. M., 2005, *A&A*, **436**, 503
Lépine S., Shara M. M., Livio M., Zurek D., 1999, *ApJ*, **522**, L121
Lery T., Baty H., Appl S., 2000, *A&A*, **355**, **1201**
Li L.-X., 2000, *ApJ*, **531**, L111
Li Z., Chiueh T., Begelman M. C., 1992, *ApJ*, **394**, **459**
Licandro J., Bellot Rubio L. R., Boehnhardt H., Casas R., Goetz B., Gomez A., Jorda L., Kidger M. R., Osip D., Sabalisck N., Santos P., Serr-Ricart M., Tozzi G. P., West R., 1998, *ApJ*, **501**, L221
Likkel L., Morris M., 1988, *ApJ*, **329**, 914
Lim A. J., Raga A. C., 1998, *MNRAS*, **298**, 871
Lin Z. Y., Weiler M., Rauer H., Ip W. H., 2007, *A&A*, **469**, 771
Lipkin Y. M., Ofek E. O., Gal-Yam A., Leibowitz E. M., Poznanski D., et al. 2004, *ApJ*, **606**, 381
Lisse C., 2002, *Earth Moon and Planets*, **90**, 497
Lister M. L., Cohen M. H., Homan D. C., Kadler M., Kellermann K. I., Kovalev Y. Y., Ros E., Savolainen T., Zensus J. A., 2009, *AJ*, **138**, 1874
Livio M., Pringle J. E., King A. R., 2003, *ApJ*, **593**, 184
Lobanov A. P., Zensus J. A., 2001, *Science*, **294**, 128
Loken C., Burns J. O., Norman M. L., Clarke D. A., 1993, *ApJ*, **417**, 515
Lopez R., Riera A., Raga A. C., Anglada G., Lopez J. A., Noriega-Crespo A., Estalella R., 1996, *MNRAS*, **282**, 470
López-Martín L., Cabrit S., Dougados C., 2003, *A&A*, **405**, L1
Lorrain P., Koutchmy S., 1996, *Solar Physics*, **165**, 115
Lovelace R. V. E., Romanova M. M., Ustyugova G. V., Koldoba A. V., 2010, *MNRAS*, **408**, 2083
Lynden-Bell D., 2006, *MNRAS*, **369**, 1167
Lynden-Bell D., Boily C., 1994, *MNRAS*, **267**, 146
Lyubarsky Y., 2009, *ApJ*, **698**, 1570
Lyubarsky Y., 2011, *Phys. Rev. E*, **83**, 016302
Lyutikov M., Lister M., 2010, *ApJ*, **722**, 197
Lyutikov M., Pariev V. I., Gabuzda D. C., 2005, *MNRAS*, **360**, 869
Macchetto F. D., 1999, *Ap&SS*, **269**, 269
MacFadyen A. I., Woosley S. E., Heger A., 2001, *ApJ*, **550**, 410
MAGIC Collaboration Albert J. and Aliu E., Aaderhub H., Barrio J. A., Bartko H., Bastieri D., Becker J. K., Bednarek W., Berger K., Bernardini E., Bigongiari C., et al. 2008, *Science*, **320**, 1752
Mannheim K., 1993, *A&A*, **269**, 67
Margon B., 1984, *ARA&A*, **22**, 507
Margon B., Stone R. P. S., Klemola A., Ford H. C., Katz J. I., Kwitter K. B., Ulrich R. K., 1979, *ApJ*, **230**, L41

Marscher A. P., Jorstad S. G., Gómez J. L., McHardy I. M., Krichbaum T. P., Agudo I., 2007, *ApJ*, **665**, 232

Marshall H. L., Harris D. E., Grimes J. P., Drake J. J., Fruscione A., Juda M., Kraft R. P., Mathur S., Murray S. S., Ogle P. M., Pease D. O., Schwartz D. A., Siemiginowska A. L., Vrtilek S. D., Wargelin B. J., 2001, *ApJ*, **549**, L167

Marti J., Rodriguez L. F., Reipurth B., 1998, *ApJ*, **502**, 337

Marti J. M. A., Mueller E., Font J. A., Ibanez J. M. A., Marquina A., 1997, *ApJ*, **479**, 151

Massaglia S., Bodo G., Ferrari A., 1996, *A&A*, **307**, 997

Matsakos T., Massaglia S., Trussoni E., Tsinganos K., Vlahakis N., Sauty C., Mignone A., 2009, *A&A*, **502**, 217

Matt S., Goodson A. P., Winglee R. M., Böhm K.-H., 2002, *ApJ*, **574**, 232

Matt S., Winglee R., Böhm K., 2003, *MNRAS*, **345**, 660

Maury A. J., André P., Li Z.-Y., 2009, *A&A*, **499**, 175

McCarthy P. J., van Breugel W., Spinrad H., Djorgovski S., 1987, *ApJ*, **321**, L29

McCaughrean M., Zinnecker H., Andersen M., Meeus G., Lodieu N., 2002, *The Messenger*, **109**, 28

McGroarty F., Ray T. P., 2004, *A&A*, **420**, 975

McGroarty F., Ray T. P., Froebrich D., 2007, *A&A*, **467**, 1197

McKinney J. C., Blandford R. D., 2009, *MNRAS*, **394**, L126

McKinney J. C., Gammie C. F., 2004, *ApJ*, **611**, 977

McLure R. J., Dunlop J. S., 2001, *MNRAS*, **327**, 199

McNamara B. R., Wise M., Nulsen P. E. J., David L. P., Sarazin C. L., Bautz M., Markevitch M., Vikhlinin A., Forman W. R., Jones C., Harris D. E., 2000, *ApJ*, **534**, L135

McNamara B. R., Nulsen P. E. J., Wise M. W., Rafferty D. A., Carilli C., Sarazin C. L., Blanton E. L., 2005, *Nature*, **433**, 45

McNamara B. R., Rohanizadegan M., Nulsen P. E. J., 2011, *ApJ*, **727**, 39

Meakin C. A., Bieging J. H., Latter W. B., Hora J. L., Tielens A. G. G. M., 2003, *ApJ*, **585**, 482

Meisenheimer K., Roser H., Hiltner P. R., Yates M. G., Longair M. S., Chini R., Perley R. A., 1989, *A&A*, **219**, 63

Meliani Z., Sauty C., Tsinganos K., Trussoni E., Cayatte V., 2010, *A&A*, **521**, A67

Mellema G., Frank A., 1997, *MNRAS*, **292**, 795

Melnikov S., Woitas J., Eislöffel J., Bacciotti F., Locatelli U., Ray T. P., 2008, *A&A*, **483**, 199

Melnikov S. Y., Eislöffel J., Bacciotti F., Woitas J., Ray T. P., 2009, *A&A*, **506**, 763

Mendoza S., Longair M. S., 2001, *MNRAS*, **324**, 149

Mignone A., Rossi P., Bodo G., Ferrari A., Massaglia S., 2010, *MNRAS*, **402**, 7

Mikołajewska J., Balega Y., Hofmann K., Weigelt G., 2010, *MNRAS*, **403**, L21

Miley G. K., Perola G. C., van der Kruit P. C., van der Laan H., 1972, *Nature*, **237**, 269

Mioduszewski A. J., Rupen M. P., Hjellming R. M., Pooley G. G., Waltman E. B., 2001, *ApJ*, **553**, 766

Mirabel I. F., Rodríguez L. F., 1994, *Nature*, **371**, 46

Mirabel I. F., Dhawan V., Chaty S., Rodriguez L. F., Marti J., Robinson C. R., Swank J., Geballe T., 1998, *A&A*, **330**, L9

Mizuno Y., Hardee P., Nishikawa K.-I., 2007, *ApJ*, **662**, 835

Mohanty S., Shu F. H., 2008, *ApJ*, **687**, 1323

Moore R. L., Cirtain J. W., Sterling A. C., Falconer D. A., 2010, *ApJ*, **720**, 757

Moraghan A., Smith M. D., Rosen A., 2008, *MNRAS*, **386**, 2091

Morganti R., Killeen N. E. B., Tadhunter C. N., 1993, *MNRAS*, **263**, 1023

Morse J. A., Hartigan P., Heathcote S., Raymond J. C., Cecil G., 1994, *ApJ*, **425**, 738

Motch C., 1998, *A&A*, **338**, L13

Mundt R., Eislöffel J., 1998, *AJ*, **116**, 860

Mundt R., Fried J. W., 1983, *ApJ*, **274**, L83

Mundt R., Ray T. P., Raga A. C., 1991, *A&A*, **252**, 740

Nakamura M., Li H., Li S., 2006, *ApJ*, **652**, 1059

Nakamura M., Li H., Li S., 2007, *ApJ*, **656**, 721

Nakamura M., Garofalo D., Meier D. L., 2010, *ApJ*, **721**, 1783

Nalewajko K., Giannios D., Begelman M. C., Uzdensky D. A., Sikora M., 2011, *MNRAS*, **413**, 333

Nichols J., Slavin J. D., 2009, *ApJ*, **699**, 902

Nichols J. S., DePasquale J., Kellogg E., Anderson C. S., Sokoloski J., Pedelty J., 2007, *ApJ*, **660**, 651

Nishizuka N., Shimizu M., Nakamura T., Otsuji K., Okamoto T. J., Katsukawa Y., Shibata K., 2008, *ApJ*, **683**, L83

Nisini B., Codella C., Giannini T., Richer J. S., 2002, *A&A*, **395**, L25

Nisini B., Bacciotti F., Giannini T., Massi F., Eislöffel J., Podio L., Ray T. P., 2005, *A&A*, **441**, 159

Noble W., 1881, *MNRAS*, **42**, 47

Noriega-Crespo A., Garnavich P. M., Raga A. C., Canto J., Boehm K.-H., 1996, *ApJ*, **462**, 804

Noriega-Crespo A., Raga A. C., Lora V., Stapelfeldt K. R., Carey S. J., 2011, *ApJ*, **732**, L16

Norman C., Silk J., 1979, *ApJ*, **228**, 197

Norman M. L., Hardee P. E., 1988, *ApJ*, **334**, 80

Norman M. L., Smarr L., Smith M. D., Wilson J. R., 1981, *ApJ*, **247**, 52

Norman M. L., Winkler K., Smarr L., Smith M. D., 1982, *A&A*, **113**, 285

Norman M. L., Winkler K. H. A., Smarr L., 1983, in A. Ferrari & A. G. Pacholczyk eds, *Astrophysical Jets*, Vol. 103 of Astrophysics and Space Science Library, Propagation and morphology of pressure-confined supersonic jets, pp. 227–250

Norman M. L., Burns J. O., Sulkanen M. E., 1988, *Nature*, **335**, 146

O'Connell B., Smith M. D., Davis C. J., Hodapp K. W., Khanzadyan T., Ray T., 2004, *A&A*, **419**, 975

O'Connor J. A., Redman M. P., Holloway A. J., Bryce M., López J. A., Meaburn J., 2000, *ApJ*, **531**, 336

O'Dea C. P., Owen F. N., 1986, *ApJ*, **301**, 841

O'Dea C. P., Daly R. A., Kharb P., Freeman K. A., Baum S. A., 2009, *A&A*, **494**, 471

Ogle P., Whysong D., Antonucci R., 2006, *ApJ*, **647**, 161

Opher M., Liewer P. C., Velli M., Bettarini L., Gombosi T. I., Manchester W., DeZeeuw D. L., Toth G., Sokolov I., 2004, *ApJ*, **611**, 575

Ostorero L., Wagner S. J., Gracia J., Ferrero E., Krichbaum T. P., Britzen S., Witzel A., Nilsson K., Ungerechts H., Vila-Vilaró B., 2006, *A&A*, **451**, 797

Ostriker E. C., Shu F. H., 1995, *ApJ*, **447**, 813

O'Sullivan S. P., Gabuzda D. C., 2009, *MNRAS*, **393**, 429

Ouyed R., Pudritz R. E., 1997, *ApJ*, **482**, 712

Ouyed R., Pudritz R. E., 1999, *MNRAS*, **309**, 233

Owen F. N., Laing R. A., 1989, *MNRAS*, **238**, 357

Owen F. N., White R. A., 1991, *MNRAS*, **249**, 164

Owen F. N., O'Dea C. P., Inoue M., Eilek J. A., 1985, *ApJ*, **294**, L85

Owen F. N., Eilek J. A., Kassim N. E., 2000, *ApJ*, **543**, 611

Owsianik I., Conway J. E., 1998, *A&A*, **337**, 69

Pakull M. W., Soria R., Motch C., 2010, *Nature*, **466**, 209

Pariat E., Antiochos S. K., DeVore C. R., 2010, *ApJ*, **714**, 1762

Peacock J. A., 1981, *MNRAS*, **196**, 135

Pearson T. J., Unwin S. C., Cohen M. H., Linfield R. P., Readhead A. C. S., Seielstad G. A., Simon R. S., Walker R. C., 1981, *Nature*, **290**, 365

Pelletier G., Pudritz R. E., 1992, *ApJ*, **394**, 117

Perley R. A., Willis A. G., Scott J. S., 1979, *Nature*, **281**, 437

Perley R. A., Dreher J. W., Cowan J. J., 1984, *ApJ*, **285**, L35

Perlman E. S., Biretta J. A., Sparks W. B., Macchetto F. D., Leahy J. P., 2001, *ApJ*, **551**, 206

Phinney E. S., 1982, *MNRAS*, **198**, 1109

Pian E., Urry C. M., Maraschi L., Madejski G., McHardy I. M., Koratkar A., Treves A., Chiappetti L., Grandi P., Hartman R. C., Kubo H., Leach C. M., Pesce J. E., Imhoff C., Thompson R., Wehrle A. E., 1999, *ApJ*, **521**, 112

Pikel'Ner S. B., 1969, *Soviet Astronomy*, **13**, 259

Piner B. G., Pant N., Edwards P. G., 2008, *ApJ*, **678**, 64

Podio L., Bacciotti F., Nisini B., Eislöffel J., Massi F., Giannini T., Ray T. P., 2006, *A&A*, **456**, 189

Porco C. C., Helfenstein P., Thomas P. C., Turtle E., McEwen A., Johnson T. V., Rathbun J., Veverka J., Wilson D., Perry J., Spitale J., Brahic A., Burns J. A., Del Genio A. D., Dones L., Murray C. D., Squyres S., 2006, *Science*, **311**, 1393

Pringle J. E., 1989, *MNRAS*, **236**, 107

Pringle J. E., 1996, *MNRAS*, **281**, 357

Pudritz R. E., Rogers C. S., Ouyed R., 2006, *MNRAS*, **365**, 1131

Punsly B., 2007, *MNRAS*, **374**, L10

Punsly B., 2011, *ApJ*, **728**, L17

Pushkarev A. B., Kovalev Y. Y., Lister M. L., Savolainen T., 2009, *A&A*, **507**, L33

Pyo T., Kobayashi N., Hayashi M., Terada H., Goto M., Takami H., Takato N., Gaessler W., Usuda T., Yamashita T., Tokunaga A. T., Hayano Y., Kamata Y., Iye M., Minowa Y., 2003, *ApJ*, **590**, 340

Pyo T., Hayashi M., Kobayashi N., Tokunaga A. T., Terada H., Takami H., Takato N., Davis C. J., Takami M., Hayashi S. S., Gässler W., Oya S., Hayano Y., Kamata Y., Minowa Y., Iye M., Usuda T., Nishikawa T., Nedachi K., 2006, *ApJ*, **649**, 836

Raga A. C., Canto J., 1995, *Rev. Mex A&A*, **31**, 51

Raga A. C., Binette L., Canto J., Calvet N., 1990, *ApJ*, **364**, 601

Raga A. C., de Gouveia Dal Pino E. M., Noriega-Crespo A., Mininni P. D., Velázquez P. F., 2002a, *A&A*, **392**, 267

Raga A. C., Velázquez P. F., Cantó J., Masciadri E., 2002b, *A&A*, **395**, 647

Raga A. C., Esquivel A., Velázquez P. F., Cantó J., Haro-Corzo S., Riera A., Rodríguez-González A., 2009, *ApJ*, **707**, L6

Raiteri C. M., Villata M., Capetti A., Aller M. F., Uckert K., Umana G., Valcheva A., Volvach A., 2009, *A&A*, **507**, 769

Raiteri C. M., Villata M., Bruschini L., Capetti A., Kurtanidze O. M., Larionov V. M., Romano P., Vercellone S., Agudo I., Aller H. D., Aller M. F., Troitsky I. S., Umana G., 2010, *A&A*, 524, A43

Rawlings S., Jarvis M. J., 2004, *MNRAS*, **355**, L9

Ray T. P., Mundt R., Dyson J. E., Falle S. A. E. G., Raga A. C., 1996, *ApJ*, **468**, L103

Rayburn D. R., 1977, *MNRAS*, **179**, 603

Rees M. J., 1966, *Nature*, **211**, 468

Rees M. J., 1971, *Nature*, **229**, 312

Rees M. J., 1978, *MNRAS*, **184**, 61P

Rees M. J., Begelman M. C., Blandford R. D., Phinney E. S., 1982, *Nature*, **295**, 17

Reipurth B., 1989, *Nature*, **340**, 42

Reipurth B., Raga A. C., Heathcote S., 1996, *A&A*, **311**, 989

Reipurth B., Bally J., Devine D., 1997a, *AJ*, **114**, **2708**

Reipurth B., Hartigan P., Heathcote S., Morse J. A., Bally J., 1997b, *AJ*, **114**, 757

Reipurth B., Bally J., Fesen R. A., Devine, D., 1998, *Nature*, **396**, 343

Reipurth B., Heathcote S., Morse J., Hartigan P., Bally J., 2002, *AJ*, **123**, 362

Reipurth B., Rodríguez L. F., Anglada G., Bally J., 2004, *AJ*, **127**, 1736

Reynolds C. S., Fabian A. C., Celotti A., Rees M. J., 1996, *MNRAS*, **283**, 873

Reynolds C. S., Heinz S., Begelman M. C., 2001, *ApJ*, **549**, L179

Reynolds S. P., 1982, *ApJ*, **256**, 13

Reynolds S. P., 1986, *ApJ*, **304**, 713

Rhoads J. E., 1997, *ApJ*, **487**, L1.

Riera A., Binette L., Raga A. C., 2006, *A&A*, **455**, 203

Rigler M. A., Stockton A., Lilly S. J., Hammer F., Le Fevre O., 1992, *ApJ*, **385**, 61

Rodríguez L. F., Mirabel I. F., 1999, *ApJ*, **511**, **398**

Rodríguez L. F., Moran J. M., Franco-Hernández R., Garay G., Brooks K. J., Mardones D., 2008, *AJ*, **135**, 2370

Romanova M. M., Ustyugova G. V., Koldoba A. V., Lovelace R. V. E., 2009, *MNRAS*, **399**, 1802

Romero G. E., Chajet L., Abraham Z., Fan J. H., 2000, *A&A*, **360**, 57

Rosen A., Hardee P. E., 2000, *ApJ*, **542**, 750

Rosen A., Smith M. D., 2004a, *MNRAS*, **347**, 1097

Rosen A., Smith M. D., 2004b, *A&A*, **413**, 593

Rudnick L., Owen F. N., 1976, *ApJ*, **203**, L107

Russell D. M., Fender R. P., Gallo E., Kaiser C. R., 2007, *MNRAS*, **376**, 1341

Ryle M., Windram M. D., 1968, *MNRAS*, **138**, 1

Sadun A. C., Hayes J. J. E., 1993, *PASP*, **105**, 379

Sahai R., te Lintel Hekkert P., Morris M., Zijlstra A., Likkel L., 1999, *ApJ*, **514**, L115

Sahai R., Kastner J. H., Frank A., Morris M., Blackman E. G., 2003, *ApJ*, **599**, L87

Sahai R., Le Mignant D., Sánchez Contreras C., Campbell R. D., Chaffee F. H., 2005, *ApJ*, **622**, L53

Sakurai T., 1985, *A&A*, **152**, 121

Salvaterra R., Della Valle M., Campana S., Chincarini G., Covino S., D'Avanzo P., Fernández-Soto A., Guidorzi C., Mannucci F., Margutti R., Thöne C. C., Antonelli L. A., Barthelmy S. D., de Pasquale M., Testa V., 2009, *Nature*, **461**, 1258

Sambruna R. M., 2007, *Ap&SS*, **311**, 241

Sari R., Piran T., Halpern J. P., 1999, *ApJ*, **519**, L17

Saslaw W. C., Valtonen M. J., Aarseth S. J., 1974, *ApJ*, **190**, 253

Saur J., Schilling N., Neubauer F. M., Strobel D. F., Simon S., Dougherty M. K., Russell C. T., Pappalardo R. T., 2008, *Geophysical Research Letters*, **35**, 20105

Sauty C., Trussoni E., Tsinganos K., 2002, *A&A*, **389**, 1068
Savolainen T., Wiik K., Valtaoja E., Tornikoski M., 2006, *A&A*, **446**, 71
Saxton C. J., Sutherland R. S., Bicknell G. V., Blanchet G. F., Wagner S. J., 2002, *A&A*, **393**, 765
Saxton C. J., Bicknell G. V., Sutherland R. S., Midgley S., 2005, *MNRAS*, **359**, 781
Saxton C. J., Wu K., Korunoska S., Lee K.-G., Lee K.-Y., Beddows N., 2010, *MNRAS*, **405**, 1816
Scheuer P. A. G., 1974, *MNRAS*, **166**, 513
Schleicher D. G., Woodney L. M., 2003, *Icarus*, **162**, 190
Schmidtke P. C., Cowley A. P., Taylor V. A., Crampton D., Hutchings J. B., 2000, *AJ*, **120**, 935
Schwarz R. A., Edge A. C., Voges W., Boehringer H., Ebeling H., Briel U. G., 1992, *A&A*, **256**, L11
Sekanina Z., Brownlee D. E., Economou T. E., Tuzzolino A. J., Green S. F., 2004, *Science*, **304**, 1769
Shaffer D. B., Cohen M. H., Jauncey D. L., Kellermann K. I., 1972, *ApJ*, **173**, L147
Shahbaz T., Livio M., Southwell K. A., Charles P. A., 1997, *ApJ*, **484**, L59
Shang H., Glassgold A. E., Shu F. H., Lizano S., 2002, *ApJ*, **564**, 853
Shang H., Lizano S., Glassgold A., Shu F., 2004, *ApJ*, **612**, L69
Shibata K., Uchida Y., 1985, *PASJ*, **37**, 31
Shibata K., Ishido Y., Acton L. W., Strong K. T., Hirayama T., Uchida Y., McAllister A. H., Matsumoto R., Tsuneta S., Shimizu T., Hara H., Sakurai T., Ichimoto K., Nishino Y., Ogawara Y., 1992, *PASJ*, **44**, L173
Shimojo M., Shibata K., 2000, *ApJ*, **542**, 1100
Shimojo M., Shibata K., Yokoyama T., Hori K., 2001, *ApJ*, **550**, 1051
Shklovskii I. S., 1963, *Soviet Astronomy*, **6**, 465
Shu F. H., 1991, *Physics of Astrophysics*, Vol. I. (Herndon, VA: University Science Books)
Shu F. H., 1992, *Physics of Astrophysics*, Vol. II. (Herndon, VA: University Science Books)
Shu F. H., Lizano S., Ruden S. P., Najita J., 1988, *ApJ*, **328**, L19
Shu F., Najita J., Ostriker E., Wilkin F., Ruden S., Lizano S., 1994, *ApJ*, **429**, 781
Shu F. H., Najita J., Ruden S. P., Lizano S., 1994, *ApJ*, **429**, 797
Shu F. H., Najita J., Ostriker E. C., Shang H., 1995, *ApJ*, **455**, L155
Shu F. H., Shang H., 1997, in B. Reipurth & C. Bertout eds, *Herbig–Haro Flows and the Birth of Stars*, Vol. 182 of IAU Symposium, Protostellar X-rays, Jets, and Bipolar Outflows. pp. 225–239
Shu F. H., Galli D., Lizano S., Glassgold A. E., Diamond P. H., 2007, *ApJ*, **665**, 535
Siemiginowska A., Stawarz Ł., Cheung C. C., Harris D. E., Sikora M., Aldcroft T. L., Bechtold J., 2007, *ApJ*, **657**, 145
Sijbring D., de Bruyn A. G., 1998, *A&A*, **331**, 901
Sikora M., Madejski G., Moderski R., Poutanen J., 1997, *ApJ*, **484**, 108
Sikora M., Moderski R., Madejski G. M., 2008, *ApJ*, **675**, 71
Sikora M., Stawarz Ł., Moderski R., Nalewajko K., Madejski G. M., 2009, *ApJ*, **704**, 38
Skopal A., Pribulla T., Budaj J., Vittone A. A., Errico L., Wolf M., Otsuka M., Chrastina M., Mikulášek Z., 2009, *ApJ*, **690**, 1222
Skorov Y. V., Rickman H., 1995, *Planetary and Space Science*, **43**, 1587
Smith D. A., Wilson A. S., Arnaud K. A., Terashima Y., Young A. J., 2002, *ApJ*, **565**, 195
Smith M. D., 1982, *ApJ*, **259**, 522
Smith M. D., 1984, *MNRAS*, **211**, 767
Smith M. D., 1986, *MNRAS*, **223**, 57
Smith M. D., 1994a, *MNRAS*, **266**, 238
Smith M. D., 1994b, *ApJ*, **421**, 400
Smith M. D., Mac Low M., 1997, *A&A*, **326**, 801
Smith M. D., Norman C. A., 1981, *MNRAS*, **194**, 771
Smith M. D., Rosen A., 2005, *MNRAS*, **357**, 579
Smith M. D., Norman M. L., Wilson J. R., Smarr L., 1981, *Nature*, **293**, 277
Smith M. D., Smarr L., Norman M. L., Wilson J. R., 1983, *ApJ*, **264**, 432
Smith M. D., Norman M. L., Winkler K., Smarr L., 1985, *MNRAS*, **214**, 67
Smith M. D., Brand P. W. J. L., Moorhouse A., 1991, *MNRAS*, **248**, 730
Smith M. D., Suttner G., Zinnecker H., 1997, *A&A*, **320**, 325
Smith M. D., Khanzadyan T., Davis C. J., 2003, *MNRAS*, **339**, 524
Smith M. D., O'Connell B., Davis C. J., 2007, *A&A*, **466**, 565
Soderblom L. A., Boice D. C., Britt D. T., Brown R. H., Buratti B. J., Kirk R. L., Lee M., Nelson R. M., Oberst J., Sandel B. R., Stern S. A., Thomas N., Yelle R. V., 2004, *Icarus*, **167**, 4
Soker N., 1998, *ApJ*, **496**, 833

Soker N., Lasota J.-P., 2004, *A&A*, **422**, 1039
Soker N., Regev O., 2003, *A&A*, **406**, 603
Soker N., Sarazin C. L., O'Dea C. P., 1988, *ApJ*, **327**, 627
Sokoloski J. L., Kenyon S. J., 2003, *ApJ*, **584**, 1027
Sokoloski J. L., Rupen M. P., Mioduszewski A. J., 2008, *ApJ*, **685**, L137
Soldi S., Türler M., Paltani S., Aller H. D., Aller M. F., Burki G., Chernyakova M., Lähteenmäki A., McHardy
    I. M., Robson E. I., Staubert R., Tornikoski M., Walter R., Courvoisier T., 2008, *A&A*, **486**, 411
Sopka R. J., Herbig G., Michalitsianos A. G., Kafatos M., 1982, *ApJ*, **258**, L35
Southwell K. A., Livio M., Pringle J. E., 1997, *ApJ*, **478**, L29
Spergel D. N., Verde L., Peiris H. V., Komatsu E., Nolta M. R., Bennett C. L., Halpern M., Hinshaw G., Jarosik
    N., Kogut A., Limon M., Meyer S. S., Page L., Tucker G. S., Weiland J. L., Wollack E., Wright E. L., 2003,
    ApJS, **148**, 175
Staff J. E., Niebergal B. P., Ouyed R., Pudritz R. E., Cai K., 2010, *ApJ*, **722**, 1325
Stanke T., 2003, *Ap&SS*, **287**, 149
Stanke T., McCaughrean M. J., Zinnecker H., 2000, *A&A*, **355**, 639
Stapelfeldt K. R., Scoville N. Z., 1993, *ApJ*, **408**, 239
Starling R. L. C., Rol E., van der Horst A. J., Yoon S., Pal'Shin V., Ledoux C., Page K. L., Fynbo J. P. U.,
    Wiersema K., Tanvir N. R., 2009, *MNRAS*, **400**, 90
Stawarz Ł., Sikora M., Ostrowski M., 2003, *ApJ*, **597**, 186
Stawarz Ł., Aharonian F., Kataoka J., Ostrowski M., Siemiginowska A., Sikora M., 2006, *MNRAS*, **370**, 981
Steiner J. E., Oliveira A. S., Torres C. A. O., Damineli A., 2007, *A&A*, **471**, L25
Sterling A. C., 2000, *Solar Physics*, **196**, 79
Sterling A. C., Hollweg J. V., 1988, *ApJ*, **327**, 950
Stewart R. T., Caswell J. L., Haynes R. F., Nelson G. J., 1993, *MNRAS*, **261**, 593
Stirling A. M., Spencer R. E., de la Force C. J., Garrett M. A., Fender R. P., Ogley R. N., 2001, *MNRAS*, **327**,
    1273
Stocke J. T., Burns J. O., Christiansen W. A., 1985, *ApJ*, **299**, 799
Stone J. M., Hardee P. E., 2000, *ApJ*, **540**, 192
Stone J. M., Norman M. L., 1993, *ApJ*, **413**, 210
Stone J. M., Xu J., Hardee P., 1997, *ApJ*, **483**, **136**
Stute M., Sahai R., 2009, *A&A*, **498**, 209
Stute M., Camenzind M., Schmid H. M., 2005a, *A&A*, **429**, 209
Stute M., Gracia J., Camenzind M., 2005b, *A&A*, **436**, 607
Suttner G., Smith M. D., Yorke H. W., Zinnecker H., 1997, *A&A*, **318**, 595
Takeuchi A., Shibata K., 2001, *ApJ*, **546**, L73
Tanvir N. R., Fox D. B., Levan A. J., Berger E., Wiersema K., Fynbo J. P. U., Cucchiara A., Krühler T., Gehrels
    N., Bloom J. S., Greiner J., Evans P. A., Rol E., Olivares F., Hjorth J., Watson D., Westra E., Wold T., Wolf C.,
    2009, *Nature*, **461**, 1254
Tavani M., Bulgarelli A., Piano G., Sabatini S., Striani E., Evangelista Y., Trois A., Pooley G., Antonelli A.,
    Salotti L., 2009, *Nature*, **462**, 620
Tavecchio F., Ghisellini G., Ghirlanda G., Foschini L., Maraschi L., 2010, *MNRAS*, **401**, 1570
Taylor A. R., Seaquist E. R., Mattei J. A., 1986, *Nature*, **319**, 38
Teolis B. D., Perry M. E., Magee B. A., Westlake J., Waite J. H., 2010, *Journal of Geophysical Research (Space
    Physics)*, **115**, A09222
Terasranta H., Valtaoja E., 1994, *A&A*, **283**, 51
Terquem C., Eislöffel J., Papaloizou J. C. B., Nelson R. P., 1999, *ApJ*, **512**, L131
Thiel K., Koelzer G., Kohl H., 1991, *Geophysical Research Letterrs*, **18**, 281
Tingay S. J., Jauncey D. L., Preston R. A., Reynolds J. E., Meier D. L., Murphy D. W., Migenes V., Quick J.,
    Sinclair M. W., Smits D., 1995, *Nature*, **374**, 141
Toma K., Yamazaki R., Nakamura T., 2005, *ApJ*, **620**, 835
Tomov T., Kolev D., Zamanov R., Georgiev L., Antov A., 1990, *Nature*, **346**, 637
Torbett M. V., 1984, *ApJ*, **278**, 318
Torbett M. V., Gilden D. L., 1992, *A&A*, **256**, 686
Trammell S. R., Goodrich R. W., 1996, *ApJ*, **468**, L107
Trammell S. R., Goodrich R. W., 2002, *ApJ*, **579**, 688
Trussoni E., Tsinganos K., Sauty C., 1997, *A&A*, **325**, 1099
Tsinganos K., Bogovalov S., 2002, *MNRAS*, **337**, 553

Tsinganos K., Trussoni E., 1991, *A&A*, **249**, 156
Turland B. D., Scheuer P. A. G., 1976, *MNRAS*, **176**, 421
Tziotziou K., Tsiropoula G., Mein P., 2003, *A&A*, **402**, 361
Uchida Y., 1969, *PASJ*, **21**, 128
Uchida Y., Shibata K., 1984, *PASJ*, **36**, 105
Uchida Y., Shibata K., 1985, *PASJ*, **37**, 515
Ulrich M., Maraschi L., Urry C. M., 1997, *ARA&A*, **35**, 445
Unwin S. C., Cohen M. H., Biretta J. A., Pearson T. J., Seielstad G. A., Walker R. C., Simon R. S., Linfield R. P., 1985, *ApJ*, **289**, 109
Urry C. M., Padovani P., 1995, *PASP*, **107**, 803
Ustyugova G. V., Koldoba A. V., Romanova M. M., Lovelace R. V. E., 2006, *ApJ*, **646**, 304
van Breugel W., Filippenko A. V., Heckman T., Miley G., 1985, *ApJ*, **293**, 83
van den Bergh S., 1970, *ApJ*, **160**, L27
van den Heuvel E. P. J., Bhattacharya D., Nomoto K., Rappaport S. A., 1992, *A&A*, **262**, 97
Veal J. M., Snyder L. E., Wright M., Woodney L. M., Palmer P., Forster J. R., de Pater I., A'Hearn M. F., Kuan Y., 2000, *AJ*, 119, 1498
Vekstein G. E., Priest E. R., Steele D. C., 1994, *ApJS*, **92**, 111
Venkatesan T. C. A., Batuski D. J., Hanisch R. J., Burns J. O., 1994, *ApJ*, **436**, 67
Venturi T., Giovannini G., Feretti L., Comoretto G., Wehrle A. E., 1993, *ApJ*, **408**, 81
Vlahakis N., Königl A., 2001, *ApJ*, **563**, L129
Vlahakis N., Königl A., 2003, *ApJ*, **596**, 1080
Vlahakis N., Königl A., 2004, *ApJ*, **605**, 656
Völker R., Smith M. D., Suttner G., Yorke H. W., 1999, *A&A*, **343**, 953
Waite Jr. J. H., Lewis W. S., Magee B. A., Lunine J. I., McKinnon W. B., Glein C. R., Mousis O., Young D. T., Brockwell T., Westlake J., Nguyen M., Teolis B. D., Niemann H. B., McNutt R. L., Perry M., Ip W., 2009, *Nature*, **460**, 487
Waldrop M. M., 1985, *Science*, **227**, 623
Wallerstein G., Greenstein J. L., 1980, *PASP*, **92**, 275
Wang Y., Sheeley Jr. N. R., 2002, *ApJ*, **575**, 542
Wardle M., Koenigl A., 1993, *ApJ*, **410**, 218
Warell J., Lagerkvist C., Lagerros J. S. V., 1999, *A&AS*, **136**, 245
Watson M. G., Stewart G. C., King A. R., Brinkmann W., 1986, *MNRAS*, **222**, 261
Wehrle A. E., Pian E., Urry C. M., Maraschi L., McHardy I. M., Lawson A. J., Ghisellini G., Hartman R. C., Madejski G. M., Makino F., Turcotte P., Unwin S. C., Valtaoja E., Villata M., Xu W., Yamashita A., Zook A., 1998, *ApJ*, **497**, 178
Wehrle A. E., Piner B. G., Unwin S. C., Zook A. C., Xu W., Marscher A. P., Teräsranta H., Valtaoja E., 2001, *ApJS*, **133**, 297
Weiler K. W., Sramek R. A., 1988, *ARA&A*, **26**, 295
Weisskopf M. C., Hester J. J., Tennant A. F., Elsner R. F., Schulz N. S., Marshall H. L., Karovska M., Nichols J. S., Swartz D. A., Kolodziejczak J. J., O'Dell S. L., 2000, *ApJ*, **536**, L81
Werner P. N., Worrall D. M., Birkinshaw M., 2001, in R. A. Laing & K. M. Blundell eds., *Particles and Fields in Radio Galaxies Conference*, Vol. 250 of Astronomical Society of the Pacific Conference Series, NGC 6251 at multiple scales and wavelengths. p. 294
Whelan E. T., Ray T. P., Davis C. J., 2004, *A&A*, **417**, 247
Whipple F. L., 1950, *ApJ*, **111**, 375
Whipple F. L., 1989, *ApJ*, **341**, 1
Whitney A. R., Shapiro I. I., Rogers A. E. E., Robertson D. S., Knight C. A., Clark T. A., Goldstein R. M., Marandino G. E., Vandenberg N. R., 1971, *Science*, **173**, 225
Wiita P. J., Siah M. J., 1986, *ApJ*, **300**, 605
Wiita P. J., Wang Z., Hooda J. S., 2002, *New AR*, **46**, 439
Williams A. G., Gull S. F., 1984, *Nature*, **310**, 33
Wilson A. S., Colbert E. J. M., 1995, *ApJ*, **438**, 62
Wilson A. S., Young A. J., Shopbell P. L., 2000, *ApJ*, **544**, L27
Wilson A. S., Young A. J., Shopbell P. L., 2001, *ApJ*, **547**, 740
Wilson H. C., 1910, *Popular Astronomy*, **18**, 477
Wilson M. J., 1987, *MNRAS*, **226**, 447
Wiseman J., Wootten A., Zinnecker H., McCaughrean M., 2001, *ApJ*, **550**, L87

Woitas J., Ray T. P., Bacciotti F., Davis C. J., Eislöffel J., 2002, *ApJ*, **580**, 336

Woitas J., Bacciotti F., Ray T. P., Marconi A., Coffey D., Eislöffel J., 2005, *A&A*, **432**, 149

Woosley S. E., 1993, *ApJ*, **405**, 273

Wu Q., Cao X., 2008, *ApJ*, **687**, 156

Xu C., O'Dea C. P., Biretta J. A., 1999, *AJ*, **117**, **2626**

Xu H., Li H., Collins D., Li S., Norman M. L., 2008, *ApJ*, **681**, L61

Xu J., Hardee P. E., Stone J. M., 2000, *ApJ*, **543**, **161**

Yamauchi Y., Moore R. L., Suess S. T., Wang H., Sakurai T., 2004, *ApJ*, **605**, 511

Yamazaki R., Ioka K., Nakamura T., 2004, *ApJ*, **607**, L103

Yelle R. V., Soderblom L. A., Jokipii J. R., 2004, *Icarus*, **167**, 30

Yirak K., Frank A., Cunningham A., Mitran S., 2008, *ApJ*, **672**, 996

Yokoyama T., Shibata K., 1995, *Nature*, **375**, 42

Zavala R. T., Taylor G. B., 2004, *ApJ*, **612**, 749

Zhang B.-B., Zhang B., Liang E.-W., Fan Y.-Z., Wu X.-F., Pe'er A., Maxham A., Gao H., Dong Y.-M., 2011, *ApJ*, **730**, 141

Zhang H.-M., Koide S., Sakai J.-I., 1999, *PASJ*, **51**, 221

Zhu Z., Hartmann L., Gammie C., 2010, *ApJ*, **713**, **1143**

Zinnecker H., Yorke H. W., 2007, *ARA&A*, **45**, 481

Zinnecker H., McCaughrean M. J., Rayner J. T., 1998, *Nature*, **394**, 862

Zirbel E. L., 1996, *ApJ*, **473**, 144

Index

Printed in the United States
by Baker & Taylor Publisher Services